MW00723397

Teaching Science to English Language Learners

"Developing the skills to integrate the teaching of science content, science practices, and language is a complex process. The chapter authors assembled in this book provide a variety of avenues for helping in-service and pre-service acquire the sophisticated knowledge and practical skills for doing so. The authors represent several theoretical stances and work within differing state mandates. However, they all afford the reader with a framework for providing intellectually stimulating science while address multiple issues of language acquisition. I found the book to be timely and an important contribution to the science education community."

–Molly Weinburgh, *William L. & Betty F. Adams Chair of Education, Director, Andrews Institute of Mathematics. & Science Education, Texas Christian University*

Luciana C. de Oliveira • Kristen Campbell Wilcox
Editors

Teaching Science to English Language Learners

Preparing Pre-Service and In-Service Teachers

Editors
Luciana C. de Oliveira
Department of Teaching and Learning
University of Miami
Coral Gables, FL, USA

Kristen Campbell Wilcox
Department of Educational Theory and Practice
University of Albany, School of Education
Albany, NY, USA

ISBN 978-3-319-53593-7
ISBN 978-3-319-53594-4 (eBook)
DOI 10.1007/978-3-319-53594-4

Library of Congress Control Number: 2017948791

© The Editor(s) (if applicable) and The Author(s) 2017
This work is subject to copyright. All rights are solely and exclusively licensed by the Publisher, whether the whole or part of the material is concerned, specifically the rights of translation, reprinting, reuse of illustrations, recitation, broadcasting, reproduction on microfilms or in any other physical way, and transmission or information storage and retrieval, electronic adaptation, computer software, or by similar or dissimilar methodology now known or hereafter developed.
The use of general descriptive names, registered names, trademarks, service marks, etc. in this publication does not imply, even in the absence of a specific statement, that such names are exempt from the relevant protective laws and regulations and therefore free for general use.
The publisher, the authors and the editors are safe to assume that the advice and information in this book are believed to be true and accurate at the date of publication. Neither the publisher nor the authors or the editors give a warranty, express or implied, with respect to the material contained herein or for any errors or omissions that may have been made. The publisher remains neutral with regard to jurisdictional claims in published maps and institutional affiliations.

Cover image © Tetra Images / Alamy Stock Photo

Printed on acid-free paper

This Palgrave Macmillan imprint is published by Springer Nature
The registered company is Springer International Publishing AG
The registered company address is: Gewerbestrasse 11, 6330 Cham, Switzerland

Acknowledgments

We would like to thank the production team at Palgrave Macmillan for the support throughout the publishing process. We also would like to recognize the support of our talented doctoral research assistants Carolina Rossato de Almeida, Sharon Smith, and Fang (Lisa) Yu. Thanks also go to our dedicated authors who made deadlines and provided excellent contributions to this volume.

Contents

List of Figures

List of Tables

1

Introduction

Luciana C. de Oliveira
and Kristen Campbell Wilcox

In the past five years, nearly every state has seen an increase in English language learner (ELL) enrollment (Cheuk 2016). During the 2012–2013 academic year, ELLs numbered 4.4 million and constituted nearly 10% of all U.S. public school students (National Center for Educational Statistics 2015). By the year 2025, ELLs are predicted to make up 25% of the student population (National Education Association 2005), with the largest number of these students found in California, Florida, Illinois, New Mexico, New York, Puerto Rico, and Texas. However, significant changes in the student population have already occurred in many states, with Arkansas,

L.C. de Oliveira (✉)
Department of Teaching and Learning, University of Miami, Coral Gables, Florida, USA
e-mail: ludeoliveira@miami.edu

K.C. Wilcox
Department of Educational Theory and Practice, University of Albany, Albany, New York, USA
e-mail: kwilcox1@albany.edu

© The Author(s) 2017
L.C. de Oliveira, K. Campbell Wilcox (eds.), *Teaching Science to English Language Learners*, DOI 10.1007/978-3-319-53594-4_1

Alabama, Colorado, Delaware, Georgia, Indiana, Kentucky, Nebraska, North Carolina, South Carolina, Tennessee, Vermont, and Virginia experiencing more than 200% growth in the numbers of ELLs in schools (NCES 2015). Given the increasing numbers of ELLs in our schools, the need for all teachers to understand their linguistic and academic needs is essential to optimize their opportunities to learn.

The Common Core State Standards (CCSS; National Governors Association Center for Best Practices and Council of Chief State School Officers 2010) set forth challenging targets for content-area teachers by integrating literacy standards in all core content areas including science. The CCSS emphasize that content-area literacy must be a high priority for content teachers. The Next Generation Science Standards (NGSS; NGSS Lead States 2013) also emphasize developing literacy skills such as articulating evidence-based scientific arguments while effectively communicating with peers and evaluating the scientific validity of the arguments to build scientific knowledge.

Language and literacy skills are critical to building knowledge in science (NGSS Lead States 2013) as students are increasingly asked to read complex informational texts and to develop sophisticated means of expression. Developing language and literacy skills in the context of content-area instruction is particularly important for the growing population of ELLs whose academic English language skills may interfere with their academic performance across subject areas (Morgan et al. 2016).

For many ELLs, one of the most challenging content areas is science, as in science students need to develop scientific reasoning skills and visual literacy as well as learn the lexical and grammatical aspects of the language of science. Nonverbal forms of representation such as graphs, diagrams, visuals, and tables are common in scientific texts and require attention from teachers. In the context of the recently published NGSS and the CCSS for Literacy highlighting the importance of developing discipline-specific reading and writing competencies in subject areas such as science, in this book, authors provide cases of science teachers' work to engage ELLs in science learning at the secondary level.

As the chapters in this book show, while science teacher educators recognize the importance of educating pre-service and practicing teachers to teach ELLs, they strive to find ways to approach the teaching of ELLs in science that will be useful for current and future teachers. The increasing diversity and established needs of today's student population, however, require science teacher educators to consciously consider what is needed to prepare science teachers for ELL students. This consideration is not enough. Rather, science teacher educators must address questions such as: What does science teacher preparation look like when it meaningfully incorporates preparation to teach ELLs? How does science curriculum change when teacher educators address ELL learning, specifically? What aspects of science teaching and learning are most relevant for the teaching of ELLs? How can ELLs achieve content and language learning in the context of various contexts? These are some questions that this book addresses.

Lourdes Cardozo-Gaibisso, Martha Allexsaht-Snider, and Cory Buxton open the book with a chapter that describes the Curriculum in Motion (CIM) model of instruction for ELLs. CIM focuses attention on (1) scaffolding scientific concepts and negotiating science talk, (2) using multimodal features to expand understandings of science talk, and (3) including students' full linguistic repertoires through translanguaging practices in the classroom.

In Chapter 3, Samina Hadi-Tabassum and Emily Reardon provide three different curriculum and instruction approaches for ELLs. One approach is to develop curriculum designed around overarching concepts and/or "big ideas" in science, especially the cross-cutting concepts from the NGSS. Another is to build a literal bridge between the home language and the language of science through a physical cognate tree that grows in the classroom. Lastly, the chapter demonstrates how to modify and adapt science texts for ELLs.

Chapter 4, by Carrie L. McDermott and Andrea Honigsfeld, describes inquiry-based instructional practices embedded in a co-taught science course focused on the elements of Physical Science and Engineering in the areas of motion, force, and stability. The core pedagogical ideas highlight project-based, student-centered integrated instruction within a collaborative service delivery model. According to the NGSS, all students are expected to engage in scientific learning while developing, strengthening, and using language, literacy, and communication skills. To meet these goals, teachers of ELLs engage their students in inquiry-based instruction and learning. A collaborative approach to science course delivery is showcased while discussing how a science/ ESOL specialist and a special education teacher addressed the aspects of NGSS's Framework including scientific and engineering practices, cross-cutting concepts, and disciplinary core ideas.

In Chapter 5, Clara Lee Brown and Mehmet Aydeniz discuss lessons learned from a state-funded professional development (PD) program designed to increase high school science teachers' pedagogical knowledge/skills of teaching CCSS-aligned informational-texts to ELLs. Results revealed that teachers' understanding of ELLs as learners and how English-proficiency specifically affects science-learning were fundamental. Knowing what ELLs can do in science through WIDA's Can-Do descriptors made it possible for teachers to tailor close-reading instructions to ELLs' specific learning-challenges. Teachers' year-long reflections on their own teaching and peers indicated that they changed perceptions about instructional-approaches for ELLs. Teachers achieved individual conceptualizations of new knowledge by figuring out classroom-applications that made sense to them.

Gretchen Oliver's single, descriptive case study in Chapter 6 shares findings and implications for engaging ELLs in science at the secondary level through a culturally responsive teaching environment and practices. Like their urban counterparts, the demographics of suburban schools have diversified in recent years with a significant increase in the number of ELLs. Teachers of ELLs must possess a unique skill set to meet the needs of this group, as "just good teaching" is not enough. Culturally responsive teaching can influence ELL student learning in positive ways, and targeted PD can promote and enhance such an approach.

In Chapter 7, Judy Sharkey and Tina Proulx share how a writing process approach combined with content instruction, and with a commitment to recognizing and valuing the rich cultural and linguistic identities that ELLs bring to our schools and communities can take place. This chapter offers insight into students' literacy and content development processes and products serving as rich examples of what ELLs, even at beginning levels of English language proficiency, can achieve in science.

Turkan and Lopez, in Chapter 8, present an analysis of the language functions embedded in the NGSS to illustrate how culturally and linguistically valid assessment practices could be brought into the classroom to support the teaching and learning of science-specific language and content.

Chapter 9, by Ana Lado and Adrienne Wright, provides examples of the use of Communicative Language Teaching (CLT) strategies in a High Intensity Language Training Extension program (HILTEX). They describe the three-step process they use to scaffold the learning of science content moving from whole group, to small group, to whole group and highlight ways that the CLT strategies used impact ELLs' confidence in their science skills, their ability to access academic texts, and ability to express their science knowledge in English.

Finally, in Chapter 10, Yuliya Ardasheva offers insight into how a "balanced" vocabulary building strategy was implemented and to what effect on ELLs. The balancing was made between both general academic vocabulary and "small" everyday words as well as science-specific, technical words and phrases both in "incidental" and "embedded" ways.

References

Cheuk, T. (2016). Discourse practices in the new standards: The role of argumentation in common core- era next generation science standards classrooms for english language learners. *Electronic Journal of Science Education, 20*(3), 92–111.

Morgan, P. L., Farkas, G., Hillemeier, M. M., & Maczuga, S. (2016). Science achievement gaps begin very early, persist, and are largely explained by modifiable factors. *Educational Researcher, 45*(1), 18–35.

National Center for Education Statistics. (2015). *Fast Facts: English Language Learners*. Institute of Education Sciences. U.S. Department of Education. Retrieved from https://nces.ed.gov/fastfacts/display.asp?id=96

National Education Association (2005). Research talking points on English language learners. Retrieved from http://www.nea.org/home/13598.htm

National Governors Association Center for Best Practices & Council of Chief State School Officers. (2010). *Common core state standards for English language arts and literacy in history/social studies, science, and technical subjects.*

Washington, DC: Authors. Retrieved from http://www.corestandards.org/assets/CCSSI_ELA%20Standards.pdf.

NGSS Lead States. (2013). *Next generation science standards: For states, by states.* Washington, DC: The National Academies Press. Retrieved from http://www.nextgenscience.org/sites/ngss/files/NGSS%20DCI%20Combined%2011.6.13.pdf.

2

Curriculum in Motion for English Language Learners in Science: Teachers Supporting Newcomer Unaccompanied Youth

Lourdes Cardozo-Gaibisso,
Martha Allexsaht-Snider and Cory A. Buxton

Things have changed, but our memories are still fresh. That Tuesday morning in October found us, a group of science and English as a second language bilingual teacher educators, inside a cold, humid, artificially illuminated trailer, removed from the main school building. More than 40 high school-aged immigrant learners were gathered together in the classroom with their science/ESOL teacher and were staring back at us curiously. After all, we were a new set of strangers coming to observe them, talk to them, and possibly teach them. Their eyes opened wide as they heard familiar sounds when we

L. Cardozo-Gaibisso (✉)
Department of Language and Literacy Education, University of Georgia, Georgia, USA
e-mail: lourdes@uga.edu

M. Allexsaht-Snider · C.A. Buxton
Department of Educational Theory and Practice, University of Georgia, Georgia, USA
e-mail: marthaas@uga.edu; buxton@uga.edu

© The Author(s) 2017

L.C. de Oliveira, K. Campbell Wilcox (eds.), *Teaching Science to English Language Learners*, DOI 10.1007/978-3-319-53594-4_2

began to talk with them in Spanish: a shared realm in which we would build relationships for the next two years.

A year passed and the trailer was traded in for a "real" classroom inside the school building, where the science/ESOL teacher and bilingual paraprofessional taught an environmental science class for these students. That classroom evolved into a bilingual space, full of signs and texts that integrated Spanish and English; both languages used on the walls and in the teaching materials. For example, bilingual general academic vocabulary cards and science concept cards were on display at all times for the students to use as touchstones when communicating their ideas orally or in writing. This classroom was unique in the school, as it allowed students to negotiate meaning freely and fluidly in both Spanish and English, to express their ideas about science multimodally and to find dynamic language support with their peers. Language-rich science investigations were central to the activities of the class. Linguistic resources flowed with ease and familiarity and were embedded in the collective movement of everyday classroom life. We all were bilingual learners, dynamically emerging. We were also scientists together, coming to know the natural world through a multitude of resources and practices. We were, indeed, teachers and students in motion.

In this chapter, we portray a dynamic approach to science instruction developed by a group of science and English as a second language (ESOL) teacher educators in collaboration with a group of newcomer, Spanish-speaking, English language learners and their high school ESOL/science teacher. Teacher educators, teachers, and student participants in this project, which we call Curriculum in Motion (CiM), worked to co-construct science teaching and learning experiences that were relevant, fluid, and built on students' linguistic, cultural, and experiential strengths (Warren and Rosebery 2008; Lee and Buxton 2010). The project conceptualized curriculum as an in-progress design, rather than as a fixed set of principles, procedures, and content to be taught (Carlson et al. 2014).

Having set the stage for the CiM project, we next provide some context for the phenomenon of rising numbers of unaccompanied youth entering U.S. schools and science classrooms that occurred during the 2014–2015 school year when our project began. We continue with a brief overview of the perspectives that informed our work as we developed the CiM project, and outline a set of principles that framed our teaching in the science classroom. Next, we share some examples of student work to illustrate how students responded to the approach. We conclude with recommendations for practice in science classrooms serving newcomer unaccompanied youth and other high school aged English language learners and with implications for science and ESOL teacher education.

In 2014 the state of Georgia began receiving a growing number of unaccompanied minors of high school age (Office of Refugee Resettlement 2016). These young people came mainly from México, Honduras, Guatemala, and El Salvador, bringing with them a wide range of prior schooling experiences, English and Spanish language proficiencies, and academic knowledge. Those who arrived in the Coral Spring School District encountered academic and linguistic challenges and opportunities as the district hastily created a pilot program called RiseUP for them at a charter school, based on a model of sheltered instruction utilizing a transitional bilingual translation approach (Harklau 1994; Short 2006). The program initially provided classes in health and life skills, English for Speakers of Other Languages (ESOL), and science, taught by English-speaking teachers assisted by bilingual paraprofessionals.

The teachers and the bilingual paraprofessionals in the RiseUP Program, who had been quickly hired after the start of the school year, also encountered challenges and opportunities. Unlike other states with large and long-standing populations of immigrants, such as California, Texas, New York, Florida, and Illinois (Office of English Language Acquisition [OELA] 2008), the arrival of a sizable group of secondary-aged newcomers was a new phenomenon in the state of Georgia in 2014. The science/ESOL teacher, health teacher, math teacher, ESOL teachers, and bilingual paraprofessionals assigned to the RiseUP program had little or no prior experience with curriculum design and educational approaches specifically designed for older English language learners (ELLs) who had experienced interrupted schooling. The RiseUP program administrator reached out to a group of educators from a local university who had been working for several years on a professional learning program to support middle and high school science and ESOL teachers in the Coral Springs School District in teaching science to English language learners.

In this uncharted territory, many shifts and changes occurred to the RiseUP program as it unfolded over four semesters. Such modifications are not unusual for programs serving newcomers, as immigrant students in these settings often encounter loosely planned school initiatives and a lack of trained professionals (Hamann and Harklau 2010). Given this dynamic and complex schooling context, as teacher educators collaborating with local teachers in the RiseUP program, we found it necessary to rethink and redesign teaching practices and create professional learning spaces, not only for a group of English language learners, but also for us as their teachers.

We found that the science and ESOL teachers, as well as the paraprofessionals, in the RiseUP program shared with our group of university educators an interest in gaining a deeper understanding of the newcomer students we were teaching and the pedagogical approaches that would support their learning. We wanted to better understand where our students came from, what content-area literacy meant for them in the particular contexts where they were studying, and the linguistic resources they brought to the classroom. As a consequence, we worked together to develop and try different strategies to co-construct language-rich science learning opportunities and spaces for the students. In the following section, we provide a brief description of some of the resources and ideas that helped us build our work, and continue to challenge our thinking in and out of the classroom. As we worked together over the two years of the project, we continued to identify perspectives from other educators and researchers that informed our articulation of a set of principles to guide our development of instructional approaches and materials in the CiM project.

Perspectives that Informed our Work in the CiM Project

In spite of the growing evidence that shows how English-only models of instruction hinder immigrant students' academic outcomes, recent educational policy tendencies have maintained and even reinforced such models of instruction (García and Kleifgen 2010). As populations of immigrant students increase in schools across the U.S., it is becoming clear that we need to prepare all teachers, not just ESOL teachers, to implement linguistically rich classroom instruction to support their English language learners in reaching high academic goals. All students, including English language learners, need support to develop college and career aspirations and readiness as well as the STEM knowledge and skills required of a twenty-first century workforce (NGSS Lead States 2013; NRC 2011; President's Advisory Commission on Educational Excellence for Hispanics 2014). As a first step in moving toward this goal for newly arrived secondary immigrant students, it is crucial for teachers to assess whether the English language learners in their classroom are "newcomers to the country, have had interrupted schooling, or have been born and grew up attending schools in the country" (de Oliveira et al. 2014, p. 8). It is essential to build on that knowledge to develop dynamic and challenging learning opportunities for all English language learners.

There is ample research evidence (e.g., Lee and Buxton 2010) that all learners, regardless of their background, benefit from learning experiences that include an acknowledgement of their cultural capital and backgrounds, and that make strong connections between academic content (such as science) and students' linguistic and cultural resources. For white, English-dominant students, these cultural and linguistic connections occur regularly in U.S. schools, but many teachers struggle to recognize the value of the experiential knowledge and skills gained in other environments that non-dominant immigrant and English language learners bring to the classroom (Stoddard 2016).

Programs for Newcomer Students and Unaccompanied Youth

Newcomer and unaccompanied youth arriving in the United States bring a multiplicity of personal histories and experiences. Many newcomer students have had interrupted schooling or a lack of formal education because of the challenges in their home countries and the time and trauma involved in migration to the United States. Interrupted schooling experiences can impact students' academic performance, but most newcomer youth face more obstacles than simply acquiring academic knowledge. Indeed, they often find that "interpersonal interactions within their new cultural context may be particularly challenging" (Patel et al. 2015, p. 9), and that their uncertain refugee and immigration status and lack of supportive extended family resources mean that concerns for daily survival and negotiating a new cultural and social context are priorities. As a result, administrators, teachers, and paraprofessionals, as well as counselors and social workers, working at schools receiving newcomer youth are confronted with the task of designing creative, responsive, and multidimensional programs to serve this new and complex population of students.

According to Hunkapiller (2010) a newcomer program is "an instructional program which addresse[s] the specific needs of recent immigrant students, most often at the middle and high school level, especially for those with limited or interrupted schooling in their home countries" (p. 10). Traditionally, these programs are located separately from other schools and serve immigrant students only (Feinberg 2000). The RiseUP program, serving the group of recently arrived unaccompanied youth in the Coral Springs School District, was situated as a "school within a school" in a charter career academy that also housed several other programs.

Teaching and Learning the Language of Science

Newcomer students whose home language is other than English, and who may have undergone interrupted schooling experiences, enter content-area classes such as science with the task of simultaneously learning a second language and grade-appropriate content knowledge. The content area of science offers unique challenges and opportunities to English language learners and their teachers, as students are increasingly asked to develop and apply scientific reasoning and investigation skills that extend beyond learning science as a body of knowledge and encourage the learning of science and engineering practices (NGSS Lead States 2013). Learning to communicate about these science practices requires the gradual mastery of the lexical and grammatical aspects of the language of science (Lee et al. 2013). Once students have begun to acquire this foundation of the language of science, "reading or listening to science can get relatively easy, but before you have done so (...) much of what is said may seem to make very little sense at all" (Lemke 1990, p. 22). Similarly, Halliday (2004) argues that one of the main challenges of learning the language of science is not exclusively related to its dense technical vocabulary, but also to its highly complex lexicogrammatical structure. As a consequence, educators need to be aware that teaching science to any student population goes beyond teaching key concepts and vocabulary, and curriculum and teaching materials need to be designed to help students take advantage of all the linguistic resources and lived experiences that they bring to the science classroom. This is especially true for bilingual middle and high school English language learners.

A Pedagogy of Translanguaging

Translanguaging is an emerging and dynamic approach in sociolinguistics and applied linguistics. For decades, scholars conceptualized multiple languages as separate entities in the brain (Haugen 1956), and thus, advocated for their instructional separation to prevent language intrusion (Weinreich 1979). More recently, informed by previous work by Cummins (1979), Dijkstra et al. (1998), Hoshino and Thierry (2011), and Grosjean (1982), translanguaging practices have gained relevance for researchers and educators working with ELLs. According to Hornberger and Link (2012 "The notion of translanguaging refers broadly to how bilingual students communicate and make meaning by drawing on and intermingling linguistic features from different languages" (p. 2).

A significant principle of translanguaging as a pedagogical approach is the conceptualization of language and context as always interacting in learning environments where students are speaking and learning multiple languages. Furthermore, a pedagogy of translanguaging encourages students to use their full linguistic repertories, in a free and dynamic way, without any restrictions (Otheguy et al. 2015). Translanguaging practices encourage educators to work at the intersection of all of their students' linguistic resources. This includes home language(s); the school/classroom language(s); everyday, academic, and scientific languages; and students' backgrounds, prior knowledge, and experiences, as well. Our own research grounded in a larger study with ELLs in general education science classrooms at the middle and high school levels has shown that students' emergent multilingualism can be an asset for science learning, especially when educators embrace the academic value of students' home language (Buxton et al. 2014). Even when systematically incorporating students' multilingual resources is not feasible, ELLs may have certain advantages in learning to use the language of science, such as the use of cognates, familiarity with multiple grammatical structures, and increased tenacity in trying to understand others' emergent science meaning making (Buxton et al. 2014).

Teachers and students can think about translanguaging as a pedagogy that recognizes the richness of multiple linguistic resources and their fluidity, and also as a way of validating the linguistic richness and dynamism that Latinos and other immigrants and linguistic groups bring to American schools (García and Wei 2013, p. 12). In the succeeding sections we illustrate how notions of translanguaging informed a set of pedagogical principles we developed in the CiM project and discuss ways in which science teachers might enact and/or adapt these principles to their school and classroom contexts.

The CiM Principles for Teaching Science with Newcomer English Language Learners

In elaborating the CiM pedagogical principles, our aim was to put students' experiences and diverse linguistic resources at the center of the science teaching and learning process. In the CiM project we viewed science knowledge as co-constructed, dynamic, and in permanent interaction with students' lived experiences. Students' background knowledge and language in science was seen not only as a way of initiating science talk, but also as a key pedagogical element that could contribute to the understanding of science as an ever-present and accessible domain of knowledge, rather than as a foreign academic realm (Banks et al. 2007).

A primary aim of our work with the newcomer students in the CiM project was to make use of three language domains—everyday, general academic, and technical or scientific—in the science classroom to enhance student meaning making. We wished to guide the students in making sense of the science concepts and the language that is used to communicate those concepts, using a wide range of multimodal resources. In supporting ELLs to make sense of both the content and the language of science, the CiM project identified the following five principles:

1. **Negotiating explicit goals for academic science content and literacy learning.** This means clearly discussing shared goals, enacting a variety of classroom practices, and explicitly providing students with strategies and techniques to own science learning.
2. **Adapting notions of translanguaging to engage students in applying all of their linguistic resources for learning science.** These resources include home languages, such as Spanish, and new languages, such as students' emerging English language proficiency, but also their developing academic and scientific languages, as well as their everyday languages. In our teaching of science, there are recurring moments in which negotiation of science talk arises.

3. **Designing lessons to elicit students' diverse background knowledge and prior experiences and to link to learning of new concepts and language in science.** Students' lived experiences in diverse geographic contexts and in their travels as immigrants, as well as their work experiences prior to emigrating and after arrival in the United States, can serve as funds of knowledge in the science classroom. This knowledge serves as a stepping stone to introducing new science concepts, elaborating hypotheses, and learning to theorize from practical experiences. In other words, these experiential resources can help students to think scientifically.
4. **Integrating multimodal resources for students to utilize in advancing understanding and communication of key science concepts and science investigation practices.** Students enact science investigations using scientific equipment and objects, integrated with a range of resources, such as diagrams, peers' knowledge and experiences, videos, and kinetic, role-playing activities for making abstract concepts more concrete without diluting the science content they are being challenged to understand. As highlighted by Halliday (1978), multimodality places emphasis on how people make meaning. Through that process, we continually create choices and select particular models (meaning potential) over others. For example, teachers use science supplies, students' bodies, the natural world, and other kinesthetic resources to engage students in modeling and explaining a scientific concept.
5. **Developing paired and small group talking and writing opportunities as daily classroom routines.** Students are given authentic tasks and bilingual literacy resources to promote reasoning and communicating together and to negotiate common understandings of science concepts and technical and academic vocabulary. Paired and small group activities, such as comprehension of a scientific text, or putting a scientific principle into practice, can provide students with the opportunity to actively interact with peers, maximize interactions in terms of quality and quantity, and offer students a safe environment to express their ideas, learn to apply science concepts, and develop science literacy.

Student Responses to the Use of the CiM Strategies in the Science Classroom

Drawing on the five CiM pedagogical principles outlined earlier, we designed written bilingual materials that served as the foundation for science investigations and other lessons we taught with the RiseUP students. In the following section, drawing on student responses to some of those bilingual written materials, we introduce several of the RiseUP students we came to know over a period of two years. We discuss these students' writing in Spanish and English to illustrate the CiM approach for framing their learning of English and Spanish biliteracy skills in the context of environmental science, the language of science, and science investigation practices.

We began our work together in the fall of 2014 by facilitating a bilingual class discussion in small groups about students' goals for learning English and their overall goals for the school year related to their health, work, families, and academic endeavors. The worksheet for recording their individual goals was developed using prompts written in both English and Spanish. We use these writings about their goals to introduce three students, and then follow with examples of their writings associated with science activities and investigations. These students were selected from among the larger group who continued into the second year of the project and who gave us permission to share their work. They represent the range of students in the RiseUP program in terms of different schooling backgrounds, sociocultural contexts, ages, home countries, and gender. We present the students' writing as it appeared in Spanish and English, showing the ways in which Spanish literacy in writing was limited but present for all of the students, while English literacy in writing began to emerge slowly. We provide loose translations of the Spanish to English, recognizing that we can only infer meaning in many cases due to the emergent nature of the writing.

Andres, a 15-year-old from Guatemala, wrote in response to the prompt *Goals for English learning and materials and strategies for learning/Metas para aprender ingles y estrategias para aprender*:

> Mi meta es tengo que haprender ingles para no para buscar un trabajo mas fasil a nostros pa yo es mi meta yo tengo que hapreder ingres es es mimi

> meta cuando yo boy ha hapreder ingres y podemo buscar. [My goal is that I have to learn English to not to look for a job that is easier for us to me that is my goal that I have to learn English is is mymy goal when I will learn English and we can look.]

Karla, a 17-year-old from México, wrote in response to the same prompt:

> Mi meta para aprender ingles es para cuando pida algo en una tienda o en un restauran y saber mas y para pedir con mas facilidad las cosas. Me gustaria aprender el vocabulario en igles matemti. Aprendo megor cuando escucho y escribiendo. [My goal for learning English is for when I ask for something in a store or a restaurant and know more and ask for things more easily. I would like to learn English *matmemti* vocabulary. I learn better when I listen and write.]

Abelino, another 15-year-old from Guatemala, wrote:

> Mi meta es para aprender ingles y tener amigos. My meta es el educación ficica y matematica y quero aprende ingles. my meta es terminar mi studio. yo quiero dar mi papel de aqui para bisitar mis family of Guatemala. yo me gusta hacer un reporte de futbol de porter. yo bisitar mis familiars en Guatemala y quero ir en cualquier lado. [My goal is to learn English and have friends. My goal is physical education and mathematics and I want to learn English. my goal is to finish my studies. I want to give my papers from here to visit my family of Guatemala. I like to write reports about soccer goalies. I visit my family in Guatemala and want to go anywhere.]

In the RiseUP program, we used the CiM pedagogical principles and practices with the goal of promoting students' abilities to think, speak, read, and write, in support of a sense-making process in science (Buxton et al. 2014). Students engaged with these practices during hands-on investigations of key concepts in environmental science. In all cases, materials and discussions were offered to students in both Spanish and English, and students were invited to respond in Spanish and English, and in both everyday and academic language, in ways that made sense to them and allowed them to express their evolving understandings.

In a set of lessons about controlling variables to ensure a fair test, students first explored bilingual concept cards explaining and illustrating the following four concepts: *variable, independent variable, dependent variable*, and *control variables*. For example, the concept card for *variable* included the following explanation (in both English and Spanish):

> In science, variable refers to something that: (1) You manipulate to study the result; (2) Changes as an effect of the experiment; or (3) You attempt to keep the same to ensure a fair test.

Following discussion and reading of the concepts and meanings associated with notions of variables using the concept cards, students worked together with the teacher to read a short scenario describing a science experiment. The passage was read and discussed in both Spanish and English.

> Lisa is studying how plants grow. She learns that light from the sun is very important for plant growth. Lisa also learns that in green houses other sources of light are used to grow plants. Lisa wonders if ***different colors of light can affect or influence how tall a plant grows***. Lisa wants to conduct an experiment to test this. She is thinking about comparing red, green and blue light to see how much the plant grows under each different color of light. Before she starts to do her experiment she needs to identify the variables in her experiment.

After reading and discussing the experiment scenario together, and drawing small illustrations to represent key ideas in the text both on the board and on their worksheets, students were directed to work with a partner, with the following prompt:

> Talk to your partner, and then write what you think for the following questions. We will discuss as a class.

Following are the writings of two of our focal students, written after discussing the questions and answers with a partner.

> Pregunta: En el experimento de Lisa sobre las plantas y el color de la luz ¿qué es lo que ella va a modificar o manipular (la variable independente)?

On the Spanish side of the worksheet, Karla wrote:*diferentes colores de la luz.*

> On the English side of the worksheet, she wrote: *Change imanip*
> On the Spanish side of the worksheet, Andres wrote:*la luz lampara*

In English at the bottom of the page written in Spanish, Andres wrote a set of English words that seemed to be copied from the board:

lisa's experiment
1) *independent variable,change manipulate*
2) *Dependent variable effect*
3) *control variable (not change)*

In another set of lessons taught after the students had been in the RiseUP program for a year, students explored the language of science practice that we refer to as *explaining cause and effect relationships* using an investigation about meteorology and weather instruments. Students were first engaged as a whole group with a set of general academic vocabulary cards, written in Spanish and English, introducing key academic vocabulary that would then appear in a short text (or *language booster*), written to introduce the science investigation. The language booster titled "Meteorology: Making Sense of Weather" was read and discussed in both English and Spanish. Following discussion of the reading, students were prompted to talk about the following question with a partner:

> *¿Por qué puede ser que tú y tu familia estén interesados en saber lo que los meteorólogos reportan todos los días sobre el clima?.* [Why could it be that you and your family are interested in knowing what meteorologists report every day about the weather?]

Abelino wrote: *es muy importante saber la clima porque la gente quieren zaber quisas es peligroso y la gente se parparan.* [it is very important to know about the weather because people want to know if maybe is dangerous and be prepared.]

Karla wrote on the Spanish side of the worksheet, but using both English and Spanish: *The weather is important because uno se puede oprehenir a weather muchas causas. y saber si ba aber tornado.* [**English**: *The weather is important because* one can learn many things about the weather. And know if there will be a tornado. **Spanish:** El clima es importante porque *uno se puede oprehenir* el clima *muchas causas. y saber si ba aber tornado.*]

At the top of the Spanish page of text Karla had written some notes for herself, recording key vocabulary from the weather concept cards in Spanish and English (*meteorologist: meteorologo; weather: crima*) and from the general academic vocabulary cards (*predict: predicir; system: sistema; interactions: interaction; result: resultados; depend: depender*). One of the vocabulary words (weather) appeared in English twice in the response she wrote to the question that was posed, interwoven with Spanish.

Following introductory lessons that were focused on developing understanding of key science concepts and vocabulary, students were engaged in an activity where they made an anemometer (for measuring wind speed) and explored how it worked. In a final discussion and writing activity focused on *explaining cause and effect relationships*, students were asked to articulate their understandings of how the anemometer worked, using the language of science and concepts related to cause and effect reasoning. The following text on the Spanish side of the worksheet prompted students' discussion and subsequent writing:

Describe los ***Efectos*** *observados durante tu investigación haciendo y probando el anemómetro.* [Describe the effects observed during your investigation and trying the anemometer.]

Karla wrote: *Con el aire de buelta y depende cuanto biento ase. The wind caused the cups on the anemometro to spin.* [**English:** With the air it spins and it depends of how much air there is. *The wind caused the cups on the anemometro to spin.* **Spanish:** *Con el aire de buelta y depende cuanto biento ase.* El viento hizo que las copas del *anemometro* giraran.]

Abelino wrote: *el efecto es infectar cosas o seres humanos.* [the effect is to infect things or human beings.]

In response to the second prompt, *Describe* ***la Causa*** *de los* ***Efectos*** *que fue observado durante la investigación hacienda y probando el anemómetro*, [Describe **the cause** of **the effects** that were observed during the investigation and trying the anemometer] students wrote:

Karla: *Causa es que no acia demas iado viento no se movio el anemometro.* [Cause is that there was not too much wind the anemometer did not move.]

Abelino: *The wind caused the cups on the anemometro to spin*

Following the recording of their ideas in an abbreviated table format, students were asked to elaborate their ideas, first using their own words (everyday language), and then using scientific language. Students' responses to the prompt, *Usa tus propias palabras para explicar las relaciones entre las* ***causas*** *y los* ***efectos*** *que observaste cuando probaste el anemómetro* [Use your own words to explain the relations between **the causes** and **the effects**you observed when you tried the anemometer], were:

Karla: *Cuando sali afuera y el anemómetro no se movio nomas quedo parado. que no abia suficiente aire para que girar el anomometro* [When I went out and the anemometer did not move it just was still. is that there was not enough air to make the anemometer move.]

Abelino: *Que cuando salimos a probar el anemómetro no realize nada para que no asia nada de aire* [Is that when we went out to test the anemometer I did not do anything because there was not air at all.]

In these extended excerpts from Karla's and Abelino's writings, we see evidence of each of them *incorporating everyday language drawn from their experiences and linking it to new concepts and language in science* they encountered in their investigation with the anemometer. In addition, we see examples in both students' writings of efforts to articulate their *understandings both in Spanish and English.*

Implications for Science and ELL Teacher Education

As we consider the work that we have done and the lessons we have learned through the CiM project, we believe that there are multiple implications for teacher education that can be generalized to at least some degree. These lessons connect directly to, and provide practical support for, the five CiM principles that we have outlined in this chapter. Generalizing implications based on the experiences of a single project is a tricky task. On the one hand, we must acknowledge the unique nature of a given context, at a given point in time, and with a given set of actors, opportunities, and constraints. On the other hand, research conducted in parallel with the implementation of an innovative pedagogical model may yield findings that can inform the needs of educators working in other contexts. These new contexts may require that aspects of the pedagogical model be adopted accordingly. That said, we believe that the following implications, as well as the five CiM principles, have much to offer other educators who are questioning the "business as usual" approaches that typically guide the teaching of science (or other content areas) with ELLs.

Understanding the unique needs of newcomers as distinct from ELLs who were born and grew up attending schools in the US. We were repeatedly reminded in the CiM project that newcomer students are an incredibly diverse group. For example, the students we worked with had a wider range of prior educational experiences, a greater likelihood of at least some interrupted schooling, a higher frequency of long hours of work outside of school, and a greater likelihood of varying degrees of trauma (both from immigration experience and from treatment in the U.S.), when compared to ELLs who are U.S. born. At the same time, the newcomers in CiM brought a diverse range of experiences and skills from their home countries that we could leverage to support their science and language learning.

Understanding the unique needs of secondary school ELLs as distinct from the needs of younger students. While the earlier research on teaching science to ELLs focused predominantly on the elementary grades (see Lee 2005), increasing numbers of middle and high school aged ELLs have required a rethinking of some pedagogical approaches. For example, the nature of secondary level science content is more abstract, both in terms of the concepts and the language required, meaning that some typical instructional strategies for ELLs (such as relying on concrete objects) may be less useful. Further, older students may need to see the personal value in what they are learning before they will fully engage in classroom activities that put them at risk of looking foolish in front of their peers. Finally, high school ELLs have a substantially shorter timeframe in which they can benefit from formal K-12 education, and educators need to work together with students to decide how to maximize that benefit.

Understanding the unique needs of ELLs in schools and districts that do not have a long history of, nor extensive resources for, meeting those needs. As we have noted, the CiM project took place in the Southeastern U.S. in what Wortham et al. (2002) has referred to as the new Latino diaspora. In such regions, teachers may not have considered the impact of increasing numbers of ELLs in their classrooms in proactive ways. Further, there are typically limited numbers of multilingual teachers, paraprofessionals, and instructional resources, while coherent and articulated district plans for newcomers are often lacking or sketchy. Thus, the arrival of newcomer students in these settings will require that programs such as CiM be developed on the fly to meet changing needs, and may call for the development of new partnerships.

Understanding the potential of school-university partnerships for supporting ELLs as well as the reasons that both groups may hesitate to embrace such partnerships. We have seen the importance of building and negotiating collaborative partnerships in order to better meet all students' needs. It is natural, however, that school personnel typically do the teaching of ELLs while attempts to understand and disseminate more effective approaches to this teaching are typically carried out by university researchers. There are a number of structural and institutional reasons for this division of labor, including different time commitments and different reward structures. That said, we found that our efforts to bridge this divide in the CiM project, while challenging, were beneficial for all involved. By co-designing, co-teaching, and co-researching, both school and university partners gained new skills and new understandings, while together, we better served the students in the project.

Understanding the unique opportunities that science provides as a context for supporting the use of multiple language domains. While it has long been recognized that science has the potential to provide a rich learning context for ELLs (Lee and Buxton 2010) due to the investigative and hands-on approaches that are ideally used to teach it, our work in CiM has helped to highlight the unique ways that science teaching can simultaneously support the multiple language domains of everyday, general academic, and technical or scientific language. Further, science topics can be clearly connected to real world needs and concerns that are of authentic interest to newcomer students (see implication #2), and can point toward fruitful career pathways that have a wide range of educational requirements (see implication #3).

Understanding how a translanguaging approach to classroom language use can support both newcomer students and their teachers. Finally, and perhaps most substantively, our work on the CiM project highlighted the critical need for each of us involved to leverage all of the linguistic resources at our disposal to maximize students' science and language learning. As the other implications of this work have highlighted, the challenges that high school aged newcomer ELLs face in U.S. classrooms can be daunting. At the same time, these students and their teachers bring a wide range of linguistic resources and lived experiences that can potentially be leveraged in new and powerful ways. While the business as usual approach to content-area learning for ELLs has not typically made the best use of these resources, innovative projects such as CiM have the potential to spread new models that support rigorous content and language learning that build on all the resources that students bring to science as well as to learning in the other content areas.

References

Banks, J. A., Au, K. H., Ball, A. F., Bell, P., Gordon, E. W., Gutiérrez, K. D., Heath, S. B., Lee, C. D., Lee, Y., Mahiri, J., Suad Nasir, N., Valdés, G., & Zhou, M.(2007).*Learning in and out of school in diverse environments: Life-long, life-wide, life-deep.*Seattle,WA:The LIFE Center, University of

Washington, Stanford University, and SRI International.Retrieved from http://life-slc.org/docs/Banks_etal-LIFE-Diversity-Report.pdf.

Buxton, C., Allexsaht-Snider, M., Kim, S., & Cohen, A. (2014). Potential benefits of bilingual constructed responses science assessments for emergent bilingual learners. *Double Helix, 2*(1), 1–21.

Carlson, J., Davis, E. A., & Buxton, C. (2014). *Supporting the implementation of the Next Generation Science Standards (NGSS) through research: Curriculum materials.* Retrieved from https://narst.org/ngsspapers/curriculum.cfm

Cummins, J. (1979). Cognitive/academic language proficiency, linguistic interdependence, the optimum age question and some other matters. *Working Papers on Bilingualism Toronto, 19,* 197–202.

de Oliveira, L. C., Maune, M., & Klassen, M. (2014). The Common Core State Standards in English Language Arts in the United States and Teaching English Language Learners: Focus on Writing. *L1-Educational Studies in Language and Literature, 14,* 1–13. Retrieved from doi: http://dx.doi.org/10.17239/L1ESLL-2014.01.01.

Dijkstra, T., Van Jaarsveld, H., & Ten Brinke, S. (1998). Interlingual homograph recognition: Effects of task demands and language intermixing. *Bilingualism: Language and Cognition, 1*(01), 51–66.

Feinberg, R. C. (2000). Newcomer schools: Salvation or segregated oblivion for immigrant students?. *Theory into practice, 39*(4), 220–227.

García, O., & Kleifgen, J. A. (2010). *Educating emergent bilinguals: Policies, programs, and practices for English language learners.* New York, NY: Teachers College Press.

Garcia, O. & Wei, L. (2013). *Translanguaging: Langauge, bilingualism and education.* New York: Palgrave McMillan.

Grosjean, F. (1982). *Life with two languages: An introduction to bilingualism.* Cambridge, MA: Harvard University Press.

Halliday, M. A. K. (1978). *Language as social semiotic.* Arnold: London.

Halliday, M. A. K. (2004). *The language of science.* London, UK: Continuum.

Hamann, E., & Harklau, L. (2010). Education in the new Latino diaspora. In E. G. Murillo, S. A. Jr., V. R. Trinidad Galvan, J. Sanchez Munoz, C. Martinez, & M. Machado-Casas (Eds.), *Handbook of Latinos and education: Research, theory, and practice* (pp. 157–169). New York: Lawrence Erlbaum Routledge.

Harklau, L. (1994). ESL versus Mainstream Classes: Contrasting L2 Learning Environments. *TESOL Quarterly, 28*(2), 241–272.

Haugen, E. I. (1956). *Bilingualism in the Americas: A bibliography and research guide* (Vol. 26). Gainesville, FL: University of Alabama Press.

Hornberger, N. H., & Link, H. (2012). Translanguaging and transnational literacies in multilingual classrooms: A biliteracy lens. *International Journal of Bilingual Education and Bilingualism, 15*(3), 261–278.

Hoshino, N., & Thierry, G. (2011). Language selection in bilingual word production: electrophysiological evidence for cross-language competition. *Brain research, 1371*, 100–109.

Hunkapiller, J. L. (2010). *Impact of a Newcomer Program on Secondary School Achievement of Recent Immigrant Students* (Unpublished doctoral dissertation). Texas A & M University-Commerce, Commerce, Texas.

Lee, O. (2005). Science education with English language learners: Synthesis and research agenda. *Review of Educational Research, 75*(4), 491–530.

Lee, O., Quinn, H., & Valdés, G. (2013). Science and language for English language learners in relation to Next Generation Science Standards and with implications for Common Core State Standards for English language arts and mathematics. *Educational Researcher, 42*(4), 223–233. doi: 10.3102/0013189X13480524.

Lee, O., & Buxton, C. (2010). *Diversity and equity in science education: Theory, research, and practice.* New York: Teachers College Press.

Lemke, J. L. (1990). *Talking science: Language, learning, and values.* Norwood, NJ: Ablex Publishing Corporation.

National Research Council. (2011). A framework for K-12 science education: Practices, crosscutting concepts, and core ideas. Washington DC. Retrieved from http://www.doe.in.gov/sites/default/files/science/next-generation-science-standards-framework-scienceeducation.pdf

NGSS Lead States. (2013). *Next Generation Science Standards.* Retrieved from http://www.nextgenscience.org/sites/ngss/files/NGSS%20DCI%20Combined%2011.6.13.pdf.

Office of English Language Acquisition, Language Enhancement, and Academic Achievement for Limited English Proficiency Students (OELA). (2008). *Biennial report to Congress on the implementation of the Title III State Formula Grant Program, school years 2004–06.* Retrieved from https://www2.ed.gov/about/offices/list/oela/title3biennial0406.pdf

Office of Refugee Resettlement. (2016). *Office of Refugee Resettlement Annual Report to Congress.* Retrieved on 10-6-2016 from http://www.acf.hhs.gov/orr/resource/annual-orr-reports-to-congress.

Otheguy, R., García, O., & Reid, W. (2015). Clarifying translanguaging and deconstructing named languages: A perspective from linguistics. *Applied Linguistics Review, 6*(3), 281–307.

Patel, S. G., Tabb, K. M., Strambler, M. J., & Eltareb, F. (2015). Newcomer immigrant adolescents and ambiguous discrimination the role of cognitive appraisal. *Journal of Adolescent Research, 30*(1), 7–30.

President's Advisory Commission on Educational Excellence for Hispanics. (2014). *Spring public meeting of the President's Advisory Commission.* Retrieved on 10-6-2016 from http://sites.ed.gov/hispanic-initiative/files/2014/05/Transcript-for-Commission-Meeting_April2014.pdf

Short, D. J. (2006). Content teaching and learning and language. In K. Brown (Ed.), *Encyclopedia of language & linguistics* (2nd ed.) (pp. 101–105). Oxford: Elsevie.

Stoddard, T. (2016). Promoting English language learner pedagogy in science with elementary school teachers: The ESTELL model of pre-service teacher education. In C. Buxton & M. Allexsaht-Snider (Eds.), *Supporting K-12 English language learners in science: Putting research into teaching practice* (pp. 139–154). New York: Routledge.

Warren, B., & Rosebery, A. (2008). Using everyday experience to teach science. In A. Rosebery & B. Warren (Eds.), *Teaching science to English language learners* (pp. 39–50). Arlington, VA: NSTA Press.

Weinreich, U. (1979). *Languages in contact: Findings and problems.* New York: Mouton Publishers.

Wortham, S., Murillo, E., & Hamann, E. (Eds.) (2002). *Education in the new Latino diaspora.* Westport, CT: Ablex.

Lourdes Cardozo-Gaibisso, a former Middle School teacher in her native Uruguay, is a PhD student at the University of Georgia's College of Education in the Department of Language and Literacy Education. She is Research Assistant for the NSF LISELL-B project. She serves as editor for the *Journal of Language and Literacy Education's Scholars Speak Out* (SSO) section and President of the Georgia Association of Multilingual and Multicultural Education. Her research focuses on how Latino emergent bilingual learners can successfully develop science literacy through a model of scientific inquiry and a pedagogy of translanguaging.

Martha Allexsaht-Snider completed her doctorate in Cross-cultural Education at the University of California at Santa Barbara (1991) and is Associate Professor in the Department of Educational Theory and Practice at the University of Georgia. Her research interests include family-school-community interactions in diverse settings including Latino/a communities and rural México, and professional development and equity in mathematics and science education. Currently, she is collaborating with colleagues in a four-year National Science Foundation grant titled *Language-rich Inquiry Science with English Language Learners through Biotechnology (LISELL-B)* that involves Latino/a middle and high school students, their families, and their science and ESOL teachers.

Cory Buxton is Athletic Association Professor of Education at the University of Georgia and a former high school science and ESOL teacher. His research fosters more equitable and engaging science learning opportunities for all students and especially for emergent bilingual learners. His most recent work is on creating spaces where students, parents, teachers, and researchers can engage together as co-learners while strengthening their academic relationships, their knowledge of science and engineering practices and careers, and their ownership of the language of science. His research has been funded by the National Science Foundation and the U.S. Department of Education.

3

Bridging Language and Content for English Language Learners in the Science Classroom

Samina Hadi-Tabassum and Emily Reardon

The Common Core State Standards (CCSS) (National Governors Association Center for Best Practices and Council of Chief State School Officers 2010) and the Next Generation Science Standards (NGSS) (NGSS Lead States 2013) are shifting how we teach science to English Language Learners (ELLs). The focus is now towards creating a symbiotic relationship between the teaching and learning of science content and the use of language to demonstrate an understanding of the science content. Science content and the language of science have become interdependent upon one another, raising the academic bar for all students, including ELLs. One of the most pronounced changes found in the set of new standards (CCSS and NGSS) is the attention

S. Hadi-Tabassum (✉)
Department of Curriculum and Instruction, Illinois University, Northern Illinois, USA
e-mail: shaditabassum@niu.edu

E. Reardon
Department of Curriculum and Instruction, Loyola University, Chicago, USA
e-mail: ereardon1@luc.edu; eereardon@gmail.com

© The Author(s) 2017

L.C. de Oliveira, K. Campbell Wilcox (eds.), *Teaching Science to English Language Learners*, DOI 10.1007/978-3-319-53594-4_3

given to language and literacy practices within and across the disciplines. In this chapter, we present a set of language and literacy practices for teachers in the general education science classroom: a) how to create curriculum that is concept-based, b) how to build bridges between the home language and the language of science, and c) how to adapt science texts for beginning-level ELLs.

Given the fact that there are more ELLs in our schools than ever before (National Center for Education Statistics 2015), the need for all teachers to understand the linguistic and academic needs of ELLs is essential. Moreover, the confluence of the demographic upswing and a new set of standards are causing curricular and pedagogical changes, which are highlighted in this chapter. We discuss how these new changes are driving much of the current conversations around instructional improvement in bilingual/ESL classrooms. At the same time, there is an important opportunity here by which mainstream science teachers can consider how language mediates knowledge building and meaning-making for ELLs. These new content standards with shared language and literacy practices can lead to potential collaborations between bilingual/ESL educators and content area science teachers. Together, the language teacher and the science teacher can work to produce content-aligned learning experiences and activities that "simultaneously develop grade-level conceptual understandings, academic practices, and the language required" to meet the expectations of these new standards for ELLs (Valdés et al. 2014, p. 25).

Teachers must have a clear vision of the goals of instruction and what proficiency means for the specific science content they are teaching to ELLs. They need to know the science content they teach as well as the language requirements for students to understand that science content. They need to be able to use their knowledge flexibly in practice to evaluate and adapt instructional materials, to represent the content in clear and accessible ways, to plan and conduct instruction using ESL methods, and to assess what ELL students are learning during and after the teaching cycle. In short, teachers need to develop and deploy a wide range of methods, strategies, and techniques to support the acquisition of scientific proficiency in ELLs.

The NGSS and ELLs

Prior to the introduction of NGSS,science teachers depended upon their state science standards for curriculum development, along with guidance from textbook publishers and their science kits and lab materials at the local level. The NGSS are currently changing the way science is taught in U.S. classrooms by moving away from a one-dimensional rendering of science as a content-driven discipline to a three-dimensional rendering of science: a) what is the knowledge of science, b) how did this knowledge come about, and c) why this knowledge came about (Cheuk 2016). The Disciplinary Core Ideas in NGSS are broad ideas that cut across multiple sciences and engineering disciplines such as life science; physical science; the earth and space sciences; and engineering, technology, and applications of science.

Rather than seeing science as a series of discrete disciplines, the NGSS is pushing for the concept of "big ideas" that connect the various disciplines. We often imagine a biologist working in different spaces and different realms than a physicist; however, the NGSS wants the next generation of students to see how a biologist and a physicist are working on the same scientific ideas and sharing disciplinary spaces between them. Examples of Disciplinary Core Ideas include how life moves from small molecules to larger organisms, how we evolve, the role of the Earth in the universe and our relationship with the Earth, how different forces interact with each other, and the processes of design and engineering. The multiple sciences or engineering disciplines are also connected with each other through seven cross-cutting concepts and "big ideas": a) patterns; b) cause and effect; c) scale, proportion, and quantity; d) systems and system models; e) energy and matter; f) structure and function; and g) stability and change. In comparison to state standards that often focused on discrete pieces of science knowledge, the NGSS group and organize the science knowledge in relation to larger, overarching ideas that connect together the discrete knowledge.

The new science standards provide synergy between the disciplinary core ideas, the cross-cutting concepts, and the practices inherent to the complexities of learning science (National Research Council 2012). For example, the NGSS also include a total of eight science and engineering practices to be integrated into the pedagogy, presented in Table 3.1.

The disciplinary core ideas represent important scientific knowledge to be learned in the K-12 years. The practices are what scientists and engineers use to investigate and build models and theories about the world, and the cross-cutting concepts are ways to organize the patterns and relationships across the domains of knowledge and practices (NGSS Lead States 2013). The practices should be integrated and connected together in the classroom so there is a scientific discourse that focuses on students taking an active stance in their own learning. In turn, the culture of the science classroom is shifting from treating knowledge as a fixed body of information to one that is generative and co-constructed, and the cognitive demand of this work makes the teaching and learning processes much more challenging.

The Old State Science Standards

Let's examine two test questions that were included in the 7th Grade Science Sample Book for the 2014 Illinois State Achievement Test (ISAT; Illinois State Board of Education 2014) to see the curricular change from state science standards to the NGSS national standards (Fig. 3.1).

These questions (#37 and #55) taken from the ISAT practice book on the Illinois State Board of Education website symbolize the science of the past. Students were expected to learn and memorize specific vocabulary within a domain of science, in this case plant botany (*compound, simple, pinnate, palmate, dicot*). Knowledge-based questions are why many teachers and students alike think science is challenging; even though they were "doing" science through labs in the classroom, they were nonetheless held accountable by the state to memorize discrete information that was removed from their everyday lives. Knowledge-based questions also require a greater degree of linguistic knowledge, making them inaccessible for students learning English. In reviewing a science text, students and teachers alike find each paragraph saturated with sentences that string together one fact

Table 3.1 NGSS practices, Pedagogy, and Sample activity

Name of NGSS practice	Pedagogy What does this look like in the classroom?	Sample activity
1. Asking questions and defining problems	Students ask each other questions, ask questions about what they are reading, ask questions about what they are observing, etc.	T places an ant farm on each table and asks students if they notice a pattern in ant behavior. Students pose and record questions in the notebook regarding ant behavior: How are the ants moving from top to bottom? How are their legs moving?
2. Developing and using models	Students create physical models to present abstract ideas and phenomena such as through pictures, a replica, scaled drawing, etc.	Students create a model of a neuron using clay and play dough. They use pictures from the Internet to make unique models as a group and then explain the model to the whole class.
3. Planning and carrying out investigations	Students carry out scientific investigations in small groups, as a class and individually. They learn to change variables in order to change the design of the investigation.	Students measure and weigh oysters and place an equal amount in two saline solutions. Then the temperature in one solution is increased. Students observe how the warmer water affects the mass of the oysters over time.
4. Analyzing and interpreting data	Students analyze and evaluate raw data and determine if there is a pattern in the data and determine whether more data needs to be collected.	Students analyze the data from the oyster experiment and determine whether a warmer saline solution leads to a loss of mass and whether more data needs to be collected in order to form a stronger claim.
5. Using mathematics and	Students use all types of math as tools to collect	Students build a simple astrolabe using a

(continued)

Table 3.1 (continued)

Name of NGSS practice	Pedagogy What does this look like in the classroom?	Sample activity
computational thinking	and analyze scientific data.	protractor to measure the shadows cast by various length objects throughout the day.
6. Constructing explanations and designing solutions	Students construct their own explanations to explain phenomena as well as apply standard explanations.	Using a lemon, lime, and orange (earth, moon, sun), students explain the seasons. They compare their own explanations in relation to the standard explanation offered by the teacher at the end of the lesson.
7. Engaging in argument from evidence	Students learn to defend and extend their ideas and explanations through argument and debate.	Using data collected by scientists across the globe, students argue whether climate change is real based on the different perspectives and whether it has effect.
8. Obtaining, evaluating, and communicating information	Students create a finished text to convey their ideas and persuade the audience.	Students write individual letters to the local newspaper editor in regard to the reported levels of lead in the school's water fountains and whether there should be greater concern.

after another. Students unfamiliar with content area reading must be taught precisely how to approach a text of this staccato nature (Shamsudin 2009).

Furthermore, if there are thousands of science words to memorize, then teachers get carried away with word lists and definitions in glossaries and spend less time on creating authentic learning experiences. Unless students are on a specialized track to study botany, the memorization of information may be of no use to them after a discrete point in

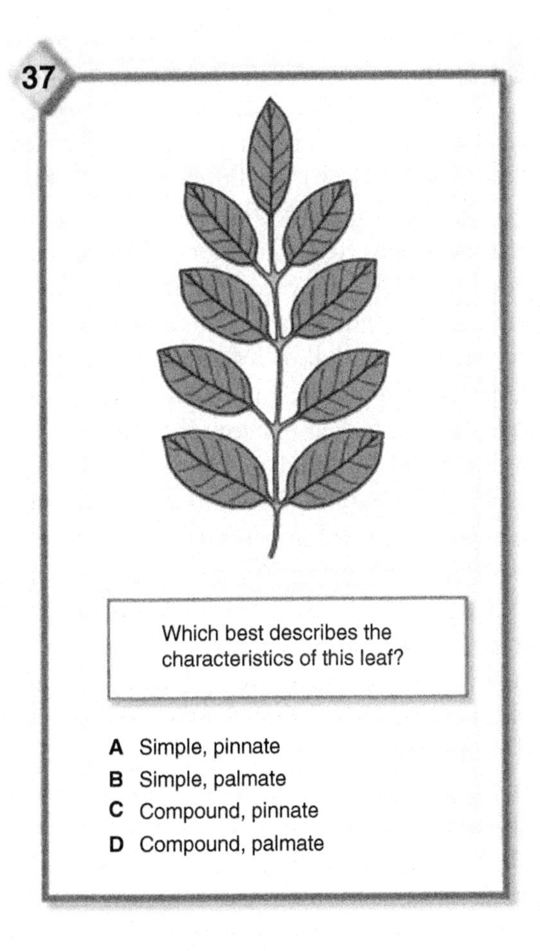

Fig. 3.1 Illinois state achievement test 7th grade science examples

time, whether it is after taking the standardized test or finishing that grade level. While learning academic vocabulary such as *palmate* and *dicot* is needed for students' academic success, teachers need to choose the words that can apply across content areas, otherwise known as Tier II words (such as *selective* and *evolutionary*), and be selective about which Tier III terms (such as *isotope* and *cerebellum*), which are subject-related words, they choose to address in each unit of study. The CCSS in

English Language Arts and Literacy have led science teachers to become much more systematic about what words to teach and how to teach them based on the Tier I, Tier II, and Tier III categories (Fig. 3.2).

The question above is also taken from the 2014 ISAT sample book for 7th grade (Illinois State Board of Education (2014). ISAT 2014); however, in Question #7 we can see how the NGSS were being transitioned into the state science test. Question #7 is still using the specific vocabulary related to botany; however, the task is asking the student to use a dichotomous key, which is a skill that can be applied across multiple subject areas. This skill of reading and interpreting a diagram to obtain information is more valuable and accessible to a wider range of students. Students are not asked to recall information about leaf structure, but they surely will be asked to follow a flow chart as a graphic organizer to attain information. By eliminating the role of memory, assessments become more reliable when the content of the assessment focuses on science comprehension and not memory retrieval.

Ideally, if a science teacher was to teach a botany unit, she could pull away from the specifics and ask herself, "what is truly important for my students to know at a conceptual level?" When learning about plants, students should be comparing the plant's structure to other living organisms and thinking about how the plant's structure, shape, and/or form affect its functioning. The students could develop a model to compare plant structures for their effectiveness in collecting water, absorbing sunlight, or playing their role as a producer in the food chain. In the process of developing and designing the model, students are accessing many more skills than when they are simply reading a text and memorizing vocabulary. They are questioning, developing and using models, collecting and analyzing data from their models, and drawing conclusions. Creating and designing the model, and even revising its prototypes, incorporates engineering and technology as well.

In continuation with the inquiry approach, when teaching cells, we teach students the classification of living things and that bacteria (prokaryotes) have no nucleus. Instead of reading about and perhaps sketching a model of different cells, science teachers can have ELLs look at models of bacteria and maybe even stained slides of bacteria. They can teach about the different shapes and groupings that make bacteria either streptococcus

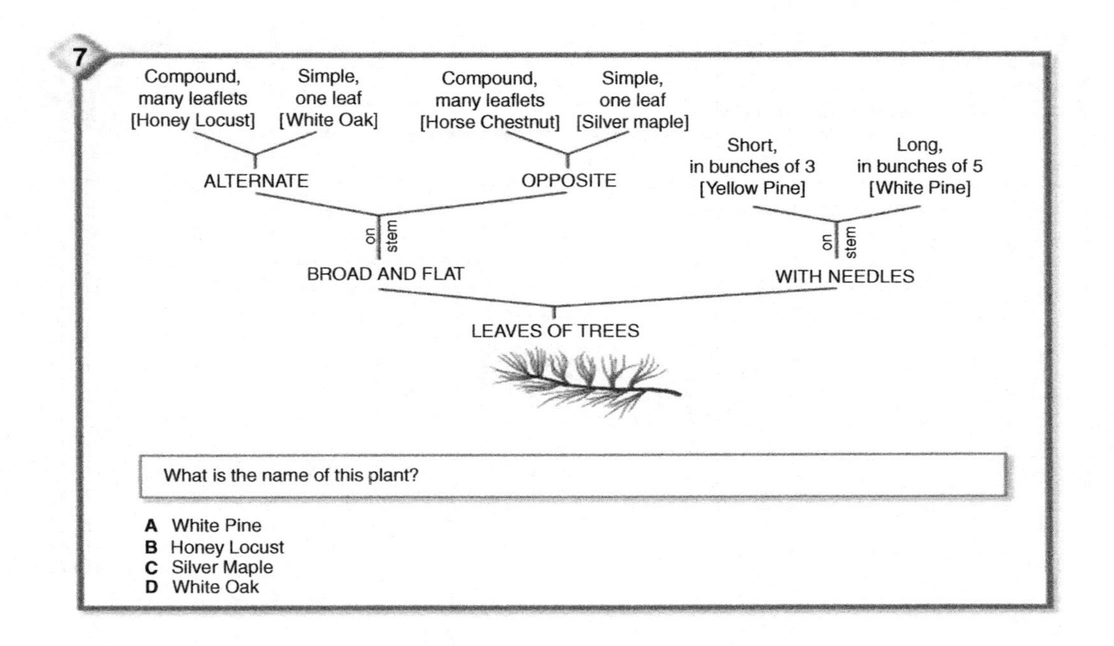

Fig. 3.2 Illinois state achievement test sample item

or bacillus. After looking at prepared slides of different types of bacteria, science teachers can then ask students to think about why and how past societies theorized about what causes people to become sick. Teachers can ask the question: "Why didn't they know to use rubber gloves, antiseptic, and wash their hands?" Students may find this dry and uninteresting. Teachers can also ask students and consider the question: "Why not think about why bacteria have no nucleus?" They can ask students to conduct their own research and develop a line of inquiry related to any topic they are interested in related to bacterial diseases. Lines of Inquiry could cover unlimited topics: the evolution of bacteria (they were here first), why/how bacteria reproduce asexually so quickly, strains of resistant bacteria (MRSA, antibiotic resistant strains, "superbugs"), how other species evolved to become more biologically sophisticated compared to bacteria, etc. These inquiry questions branch across a variety of subject areas and can be aligned with CCSS in language, math, as well as NGSS.

Concept-Based Teaching

As mentioned earlier, one of the major changes that NGSS has brought to the science classroom is a move away from micro-level, skill-based lessons that focus on memorizing discrete pieces of information to now a much broader, macro-level perspective on how scientific knowledge is connected and interdependent. Instead of teaching a typical unit on anatomy that covers all of the body systems and organs, first system-by-system and then organ-by-organ, science teachers can design instead a unit around the broader concepts of structure and function. Students would look at the structure of the human form (as well as other organisms) and consider ways in which structure is related to function. Through these epistemological investigations, students are making connections at the conceptual level instead of memorizing isolated facts and terms. The summative assessment at the end of the unit would ask students to synthesize generalizations about how structure is related to function, not only in our bodies but across all areas of science such as physical science and engineering. Table 3.2 shows potential generalizations at the end of the unit:

Table 3.2 Generalizations

Structures in anatomy	Relation to functions
Brain wrinkles	More surface area, room for neurons to travel, increases speed of delivery.
Chewing our food	Mechanical digestion allows for quicker and more efficient chemical digestion.
Small cells	Smaller cells can absorb and expel nutrients and wastes and more efficient rates than a few large cells.
Alveoli in our lungs	These allow the gas exchange in the lungs to work at a much faster and more efficient rate.

In the end, we need to develop ELLs with life-long, enduring understandings that are transferrable across disciplines. These understandings are at the conceptual level and will hopefully help them to make similar connections when learning future content. If they are late-arrivers (after 3rd or 4th grades), they have more content to catch up on than is probably feasible through simply a knowledge-based curriculum. By teaching at a conceptual level, ELLs are given access to Tier II cross-cutting terminology, such as *structure* and *function*, and are also hopefully able to learn to make connections in their learning, instead of just remembering and regurgitating isolated terms, formulas, equations, and facts.

Building a Bridge Between Language and Content

Academic Language

A common challenge that ELLs face is the specialized academic language of science, the load of academic words in science, and the speed at which words are taught, leading to the difficulty of learning science for all students. Science is a subject consumed in academic language with very specific and precise meanings, and science is seen as a language all its own (Brown and Ryoo 2008; Lee 2002; Michaels et al. 2008). Words that are known in the everyday context may also have an entirely different meaning in the context of science such as the word *properties*, which has two meanings or lexemes—one in our everyday reality (e.g., *I own many properties*) and one in science (e.g., *the properties of iron*). The meaning of *properties* would then constitute academic vocabulary when found in science texts, and therefore ELLs need to know that there are many words in English that are homonymous and have multiple meanings due to their different origins. How to use those precise meanings correctly with care and accuracy is a skill ELLs gain along their way to native-like proficiency over several years (Aitchison 2003). Over time, ELLs will learn to use specific words according to specific contexts and learn to separate out the multiple meanings.

In an average elementary science textbook, there are 20–30 new academic words introduced in a specific chapter (Meara 1980). If two chapters are read a month, then there is a potential to acquire knowledge of at least 400 plus academic words a year in science. However, within that list, there are both Tier II and Tier III words, some specific to science and others that cut across multiple disciplines. Yet, in the end, students should be held accountable to know and apply their understanding of these new academic words. If ELLs do not acquire the meaning of these words, then it can lead to misconceptions in science as well as a turn against science as a subject due to a lack of understanding (Pearson et al. 2007).

At the same time, one must question whether students should learn so many new science words and whether they are able to do so well. Jeanne Chall's (1996) seminal research points out that kids typically learn only 8–10 words per week for full mastery. Chall also used the concept of the "fourth-grade slump" to refer to the time when words become challenging for all students, because they are more abstract, academic, literary, and less common. Other researchers point out that children are acquiring knowledge of new words, many more than 10 a week, through many different explicit and implicit ways and that teachers must be much more strategic about how words are introduced in the classroom and how often they are used in classroom discourse (Baker et al. 1998). The student may not have mastered the meaning of all these new incoming words in school science, and perhaps a basic, fuzzy meaning of a word may suffice in the end for comprehension. When teaching ELLs, it is imperative to keep in mind the academic gap that might exist in relation to their monolingual peers in terms of both everyday words and science words. It may take a few years before the ELL catches up to her/his peers.

We tend to see ELLs as a monolithic whole when in fact there is a spectrum of ELLs in our classroom based on multiple variables—from language proficiency levels to educational backgrounds to depth of poverty and even immigration status (Goodwin 2002). Some ELLs may be recent arrivals from foreign countries while others have been born in the United States and have a home language other than English. Some ELLs may have acquired English listening and speaking skills but may not be as proficient in reading and writing. Others may have struggles with mostly reading. Memory may play a role in why some ELLs struggle academically. Yet, general education teachers may not see a spectrum of students in front of them; rather, they might homogenize the traits of all ELLs and see them from one perspective only—a deficit perspective in which the ELLs simply lack English.

An "additive approach" is needed in which mainstream teachers see the strengths the ELLs bring with them to the classroom such as knowledge of their home language and a grit personality that allows them to persevere against obstacles (Cummins 2000). Research has shown that states with an additive bilingual policy and programming that is strengths-based have a statistically significantly positive effect on fourth grade National Assessment of Educational Progress (NAEP) reading achievement scores among both Hispanic ELLs and Hispanic non-ELLs (Lopez and McEneaney 2012). However, when it comes to achievement in science, there is a need for more quantitative and qualitative research examining the role of bilingual education in increasing science achievement based on longitudinal, norm-referenced data (Lee 2005). With the new PARCC science test, which is currently an English-only assessment with set accommodations for ELLs, we will now be able to see nation-wide data aligned with the NGSS and analyze how students perform across the country, grade levels, and various demographics.

The Use of Cognates to Build Bridges

Yet, when it comes to acquiring academic literacy in science, one of the most effective techniques is building a bilingual bridge between the student's home language and English. In their text, *Teaching for*

Biliteracy, Beeman and Urow (2012) argue that ESL and bilingual teachers must build a literal “bridge” that brings together the student’s home language and English, that asks students to contrast the two languages and analyze how the languages are the same and different, and that ultimately asks students to transfer knowledge from one language to another. By building a bridge between the two languages, ELLs develop a heightened metalinguistic awareness (thinking about their own thinking) as they move back and forth between languages (Bialystok 2001). At the same time, the biliteracy framework argues that the foundation must be first built in the more secure home language and then transference to English comes next.

The conscious use of cognates is one common technique to bridge building. Cognates are words in two or more languages that share the same meaning but are often spelled differently and/or pronounced differently. Cognates share the same root meaning and that root meaning has traveled across multiple languages throughout human history. Cognates, however, only occur in languages that are related to one another and come from the same linguistic family. Therefore, it is not possible to use cognates for Chinese and English since they do not come from the same linguistic family. Spanish, on the other hand, has several hundred cognates with the English language due to the fact that they are both members of the Proto-Indo-European (PIE) macro language—an ancestor that connects languages stretching from Northern India (Hindi, Marathi, Gujarati) to the British Isles. The word for “star” in Hindi is स्टार and its transliteration is [sṭāra]. The root meaning of *star* eventually became *astral* when it reached the Greek language and then *star* in English. We know the *s-t-a-r* root traveled along the PIE corridor and maintained its meaning while changing its sound and spelling.

The genetic ties between Spanish and English are not as strong as German and English and Dutch and English since English is a member of the Anglo-Saxon branch of languages. On the other hand, Spanish is a daughter of the Latin language along with French and Italian. Close to 12,000 words from Latin spilled into the English language during the English Renaissance around 1500–1650 when science and math opened up the immediate world—from the inner world of anatomy to the outer world of astronomy (Finnegan 2016). Within the Latin lexicon, there were also a smaller percentage of Greek words that made their way into the English language during the Renaissance. Thus, academic word lists in the math, social studies, and science subjects will include hundreds of root words that came into English from mostly Latin and Greek.

In a science classroom, if the teacher is knowledgeable of Spanish, then the conscious teaching of cognates can occur. However, even if a teacher is not knowledgeable of Spanish, then the Spanish-speaking ELLs can take the lead and help start the bridge building process with teacher support. Before beginning the formal process of combing for cognates, it is imperative that the teacher talk about the differences between true cognates, false cognates, and partial cognates in English and Spanish. True cognates share the same meaning but may look and sound different: *nación*/nation, *importante*/important, *música*/music, etc. Partial cognates occur due to homonymous word pairs in one of the two languages; a word in Spanish or English may be a true cognate to one of the homonymous pairs but not to the second one.

For example, the Spanish word *preciso* has two meanings and only one of its two meanings is a true cognate with a word in English, but the second meaning of *preciso* does not have a true cognate in English. One meaning of *preciso* is *exact*, so it aligns well with the English word *precise* and one would say that *preciso* and *precise* are true cognates. However, there is a second meaning of *preciso* in Spanish which means *necessary*; therefore, this second meaning does not have a cognate in English: *Es preciso que hacerlo ahorita.* [It is necessary to take care of if now]. Here is another example: the Spanish word *convicción* matches with the true English cognate *conviction,* which means *one's beliefs,* but it does not match up with the penal *conviction.* The word in Spanish for a penal conviction is *condena.*

There are, however, variations in cognates depending on the various dialects of the Spanish language. The word *torta* (which means "twisting, turning, full of curves" in Latin) means *a round cake* in some Spanish-speaking countries (such as Argentina and Spain), but in other Spanish-speaking countries, the word for *cake* is *pastel,* such as Mexico where a *torta* is a sandwich and not a cake. Yet, the root *tort* is a cognate that bridges English and Spanish. Furthermore, the list of false cognates should be shared with students as well, even though the majority of false cognates are outside of the science discipline. Here are some common false cognates in which the two words may sound and look alike but do not share the same meaning: the Spanish word *carpeta* means *folder* and not the English word *carpet*; the Spanish word *fábrica* means a *factory* and not the English word *fabric*.

In a sample lesson conducted with middle school students in a science classroom, the teacher read an article about the human body and within that three-page article,the students found close to 50 true cognates: *anatomía/anatomy, esófago/esophagus, cráneo/cranium, intestine/intestine, ligamento/ligament, pulmón/pulmonary, músculo/muscle, nervio/nerve, órgano/organ, esqueleto/skeleton, estómago/stomach*,etc. After reading the text,the class as a whole identified a list of true cognates. Then each pair of students was assigned a pair of cognates.The teacher in her classroom created a cognate tree out of butcher paper. The trunk of the tree was taped onto the back of the door and brown branches stretched out from the trunk on the door and across the classroom walls. In fall, the students wrote the Spanish word on a yellow leaf and the English word on an orange leaf to match seasonal colors. Then the teacher taped the pair of cognates along a branch. The teacher also had a stack of die-cut leaves in a basket for students to take and write out their "found" cognates outside of the formal teaching of cognates. Students found cognates when doing homework, while driving along the roads on signs and billboards, and in books and materials at home. After the winter holidays, the teacher changed the colors to different shades of green to welcome spring. By the end of the school year, the cognate tree grew across the classroom and became a visual marker of bridge building between languages (Fig. 3.3).

In addition to building bridges between languages, a science teacher needs to become conscious of how language proficiency comes into play when teaching ELLs. When teaching science content, the teacher must ask her/himself what type of linguistic obstacles might prevent the ELLs from fully understanding the science content. Furthermore, lesson plans can seamlessly integrate the content knowledge objectives with the language objectives so that they are mutually reinforced and ELLs learn to transfer linguistic skills and apply them to science knowledge (DelliCarpini 2008). Here language objectives are not just vocabulary and grammar objectives; rather, they tell us what ELLs are doing in that science classroom in relation to listening, speaking, reading, and writing. At the same time,

Fig. 3.3 Cognate tree

language and content objectives must be very specific and detailed. How will ELLs transfer what they are listening to in their science classroom into a demonstration of their understanding? What will they specifically write and for how long? What will they specifically read, and how will they respond to the reading? Holbrook and Rannikmae (2009) posit that scientific literacy is dependent on the need to teach language objectives in tandem with content knowledge: (a) speaking to describe and support ideas (for example, "What do you know about ___?" "Can you describe what ___ is?" "Can you tell me, 'What does ___ mean?'"); (b) writing to remember specific information; (c) reading to expand learning and to find answers; and (d) listening to enhance thinking. The new CCSS and NGSS standards push all teachers to carefully examine the role of language in their classrooms and how language becomes a conduit to better understanding the academic content.

Modifying and Adapting Along the Way

When teaching beginning-level ELLs in a science classroom, another area of concern is the readability of the science texts and what can be done to modify and adapt the original text in order to increase student comprehension. ELLs with low levels of English proficiency have to be formally taught how to read complex texts in science, whether it is through a close reading approach or the annotation of a science text (Shamsudin 2009). However, for higher levels, ELLs need access to the language of science, as it is written, without modifications—therefore there needs to be a differentiation of modification based on the ELL's level of English language proficiency (de Oliveira and Shoffner 2016). Yet, given the fact that most school districts do not change their textbooks until 7–10 years later, due to the heavy cost where each textbook could be $100 (Rapp 2008), teachers can also create their own science texts that include more current information, and these newly created texts can be shared across the school, district, and even through social media sites—an empowering act for teachers who feel like professionals when creating and designing their own curricula that is not paper-bound.

At the same time, there are many concerns that teachers have with their current science textbooks, such as formatting, layout, chapter structure, and content representation. One concern is that science textbooks are much more challenging to navigate today than in earlier times when textbooks were more streamlined. Oftentimes, there is too much information and not all the information is truly relevant when it comes to better understanding larger concepts. The inclusion of too much supporting evidence such as graphs, tables, diagrams, photographs, drawings, symbolic representations, and other texts can also be overwhelming—along with whether there is a physical alignment between the words themselves and the supporting evidence. Not only are ELLs grappling with complex science theories and abstract ideas that are oftentimes unobservable and removed from their lived experiences, they also have to "combine, interconnect, and integrate verbal text with mathematical expressions, quantitative graphs, information tables, abstract diagrams, maps, drawings, [and] photographs" (Lemke 1998, p. 88).

In order to decrease the linguistic load but maintain the cognitive load, teachers have to "translate" the science knowledge in their textbooks for beginning-level ELLs by creating new texts and/or modifying and adapting extant text so the science knowledge is comprehensible for this group of ELLs. One method is to create a new text by modifying the language in the original science chapter, then adding more supporting photographs and drawings, and reformatting the next document so it is easy to navigate for all students. Even though creating new texts might be time consuming and take up to a few hours, it does decrease the frustration that may occur when beginning ELLs are not fully comprehending what they are reading as well as decreasing the time looking and searching for readable texts on a specific science topic. At the same time, even when the linguistic load decreases, it is imperative that the academic language in the original text stay intact and a glossary is used to scaffold the meaning of the academic words. The academic language is needed to conduct scientific inquiries, construct theoretical explanations of the natural phenomenon, and communicate scientific principles and procedures (Fang and Wei 2010).

Next is an example of a text adaptation in science. This article titled "Genome Reveals Why Giraffes Have Long Necks" was published in the journal *Nature* in March 2016 by Bethany Augliere. The introduction begins with a dizzying amount of data.

> Call it a tall task: researchers have decoded the genomes of the giraffe and its closest relative, the okapi. The sequences, published on May 17 in Nature Communications, reveal clues to the age-old mystery of how the giraffe evolved its unusually long neck and legs.
>
> Researchers in the United States and Tanzania analyzed the genetic material of two Masai giraffes (Giraffa camelopardalis tippelskirchi) from the Masai Mara National Reserve in Kenya, one at the Nashville Zoo in Tennessee and an okapi fetus (Okapia johnstoni) from the White Oak Conservation Center in Yulee, Florida. "This is one more wonderful demonstration of the power of comparative genomics to connect the evolution of animal species on this planet to molecular events that we know must underpin the extraordinary diversity of life on this planet," says David Haussler, director of the Genomics Institute at the University of California, Santa Cruz.
>
> As the tallest mammals on Earth, giraffes can reach heights up to nearly 6 metres, with necks stretching 2 metres. To prevent fainting when they lower their heads to drink water, giraffes have developed an unusually strong pumping mechanism in their hearts that can maintain a blood pressure 2.5 times greater than that of humans. To keep their balance and reach sprints up to 60 kilometres per hour, giraffes have a sloped back, long legs and short trunks. But their closest relative—the okapi—resembles a zebra, and lacks those modifications.
>
> Previous genetic research has suggested that the okapi and the giraffe diverged from a common ancestor roughly 16 million years ago, says study co-author Douglas Cavener, a biologist at Pennsylvania State University in University Park. But the latest study found that the two species diverged much more recently, about 11.5 million years ago. To identify genetic changes associated with the giraffe's unique qualities, Cavener and his colleagues compared gene-coding sequences of the giraffe genome to those of the okapi, and then to those of more than 40 other mammals, including sheep, cows and humans.

In this introduction (Augliere, 2016), there are many names, dates, and places listed, which might not be relevant for young readers. At the same time, this article is a great one to adapt since it contains contemporary information and has an appealing topic for students. The first step when modifying this article is to determine what essential understandings are relevant and which ones are not. Then with a bulleted list of essential understandings, the teacher would rewrite the text in a Word document with formatting and layout in mind (font type and font size, spacing, pagination) as well as keeping the writing style and narrative structure in place when rewriting the text. Lastly, the teacher would search the Internet and other sources in order to add supporting images and evidence.

Here is what the adapted introduction would look like in print (Fig. 3.4):

In the appendix, we have attached a checklist that teachers can use to begin adapting science texts for beginning ELLs. Not all criteria need to be met in the checklist; rather, the teacher can decide which texts need to be adapted and in what manner. In bilingual classrooms, it might be challenging finding adapted texts in the student's home language. Therefore, bilingual teachers should be encouraged to write their own bilingual science texts based on research mostly in English. One teacher decided to research on her own about ground hogs in English and then wrote her own text in Spanish to use on Ground Hog's Day or *Día de la Marmota*. Her book used images from the web, streamlined the essential information in short sentences and had only a few sentences per page, highlighted key vocabulary, and included comprehension exercises at the end of the book. After she read the book aloud, the book was laminated and placed in the library for Spanish texts. It was one of the most popular books to read.

Although text adaptations might take some time, in the end, the time is well spent, and the adaptation can be used for several years and become a part of the classroom culture. There are other examples of text adaptations that include the use of writing stems, keywords, and graphic organizers used as learning scaffolds. In essence, text adaptations help beginning-level ELLs "translate" the science knowledge that is embodied in various text representations, and in turn, use that knowledge and identify appropriate conclusions. By reading text adaptations

Giraffes are the tallest **mammals** on the planet. They can be 20 feet tall. How did giraffes get so tall and why? Why do they need necks as tall as 7 feet? When giraffes bend down to drink water, they can become faint and fall but their hearts are so strong that they pump blood to the brain and the giraffe in the end does not faint. They also have long legs and a sloped back so that they do not fall when running as fast as 37 miles per hour.

Believe it or not but the giraffe is actually related to another animal you might have seen in thezoo—the okapi. But why does the okapi look so differently from its relative the giraffe? Scientists in the United States and Tanzania, where giraffes and okapi can be found in their natural settings, are looking inside both animals and are mapping their **DNA**. The scientists are comparing and contrasting the giraffe and okapi to see what happened when they separated from an earlier ancestor almost 11.5 million years ago. That's how old the animals are!

11 Million Years Ago: The giraffe split away from the okapi.

- Scientists are mapping the DNA of the giraffe and the okapi to see why they are different and how the giraffe developed a long neck and long legs.

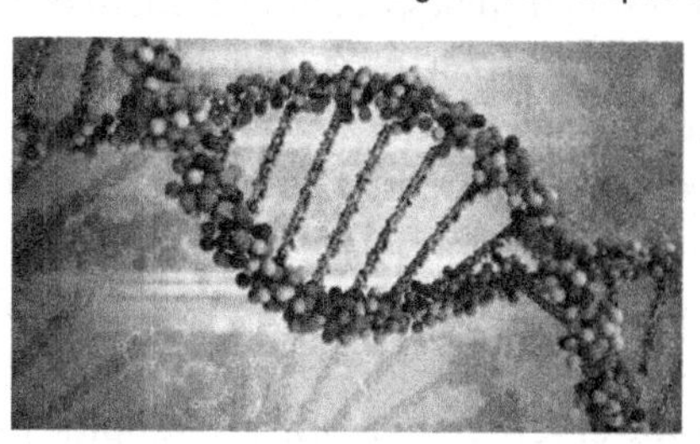

Glossary: Here is where you can look up the meaning of the science words.

Fig. 3.4 Text adaptations and visuals

Mammals: A mammal is an animal that has hair or fur on its body. A mammal drinks milk from its mother. A mammal gives live birth to babies. A mammal is also warm blooded, which means the temperature of its blood stays the same. Here are some pictures of mammals.

What is a Mammal?

DNA: All living things have DNA [known as "deoxyribonucleic acid" in science] inside of their bodies. The DNA is made up of chromosomes. The chromosomes are then made up of 4 different types of molecules [adenine, guanine, cytosine, and thymine] that connect together like a ladder and look like a ladder. Your DNA tells us about who you are on the outside and the inside.

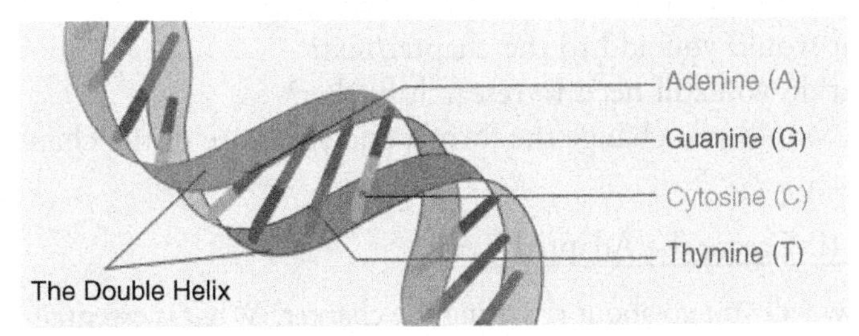

Fig. 3.4 (Continued)

in science, the linguistic load is decreased, and now time is spent instead on making sense of the science content.

At the same time, when adapting texts, it is essential for the teacher to maintain the richness, complexity, and rigor of the original text, as well as to cite the original text and its author(s) in the adaptation. The relevance of the science content must stay in tact while the linguistic load decreases. Others argue that text adaptations build too many scaffolds and that the ELL must at some point read dense, thick science text on her/his own, especially as the levels of English proficiency increase. Yet, before they reach that independent stage of reading complex science texts and making sense of such texts on their own, a teacher must explicitly differentiate the texts in order to ensure comprehension. Otherwise, a beginning ELL becomes frustrated and may give up entirely on science when the language prevents access to the content.

Appendix

Part I: Explore the Original Text

1. Do you think this science chapter/text is easy to navigate?
2. What do you think about the structure and format of the chapter/text?
3. What information do you think is essential for students to know in the chapter/text?
4. What would you remove from the chapter/text?
5. What would you add to the chapter/text?
6. What do you still need to research further?
7. How would you change the format and structure of the chapter/text?

Part II: Create the Adapted Text

1. How will you go about rewriting the chapter? What is essential to keep?
2. What academic words need to stay intact?
3. What images will you add?
4. What other supporting evidence will you add?
5. What font size and type will you use?
6. Are you using headings and subheadings?
7. Will you add questions at the end?
8. Will there be a glossary?
9. How will you space the text?
10. What is the layout of the text?
11. Is the newly created text easy to navigate for students?
12. How did you change the voice of the text so it is easily accessible to students?

Part III: Implement the Adapted Text

1. Was the adapted text as rigorous as the original?
2. Did the adapted text feel dumbed down?
3. Was the academic content comprehensible for ELLs?
4. How can the adapted text be used as a resource in the classroom?
5. How can the adapted text be shared across the grade, school, and district?

References

Aitchison, J. (2003). *Words in the mind.* Oxford, UK: Blackwell.

Augliere, B. (2016). Genome reveals why giraffes have long necks. *Nature, 59*(3), 444–445.

Baker, S., Simmons, D., & Kame'enui, E. (1998). *Vocabulary acquisition: Synthesis of the research.* Washington, DC: U.S. Department of Education, Office of Educational Research and Improvement, Educational Resources Information Center.

Beeman, K., & Urow, C. (2012). *Teaching for biliteracy. Strengthening bridges between languages.* Philadelphia, PA: Caslon Publishing.

Bialystok, E. (2001). Metalinguistic aspects of bilingual processing. *Annual Review of Applied Linguistics, 21*, 169–181.

Brown, B. A., & Ryoo, K. (2008). Teaching science as a language: A "content-first" approach to science teaching. *Journal of Research in Science Teaching, 45*(5), 529–553.

Chall, J. (1996). American reading achievement: Should we worry?. *Research in the Teaching of English, 30*(3), 303–310.

Cheuk, T. (2016). Discourse practices in the new standards: The role of argumentation in common core-era next generation science standards classrooms for English language learners. *Electronic Journal of Science Education, 20*(3), 92–111.

Cummins, J. (2000). *Language, power and pedagogy: Bilingual children in the crossfire.* Clevedon, UK: Multilingual Matters.

de Oliveira, L. C., & Shoffner, M. (Eds.) (2016). *Teaching English language arts to English language learners: Preparing pre-service and in-service teachers.* London, England: Palgrave Macmillan.

DelliCarpini, M. (2008). The success with ELLs: Working with English language learners: Looking back, moving forward. *The English Journal, 98*(1), 98–101.

Fang, Z., & Wei, Y. (2010). Improving middle school students' science literacy through reading infusion. *The Journal of Educational Research, 103*, 262–273.

Finnegan, E. (2016). *Language: Its structure and use.* Independence, KY: Wadsworth Publishing.

Goodwin, A. L. (2002). Teacher preparation and the education of immigrant children. *Education and Urban Society, 34*, 156–172.

Holbrook, J., & Rannikmae, M. (2009). The meaning of scientific literacy. *International Journal of Environmental & Science Education, 4*(3), 275–288.

Illinois State Board of Education (2014). ISAT 2014 Science Sample Book—Grade 7. Retrieved from http://www.isbe.net/assessment/pdfs/2014/isat/Gr7-SB-Science-14.pdf.

Lee, O. (2002). Science inquiry for elementary students from diverse backgrounds. In W. G. Secada (Ed.), *Review of Research in Education* (Vol. 26, pp. 23–69). Washington, DC: American Educational Research Association.

Lee, O. (2005). Science education with English language learners: Synthesis and research agenda. *Review of Educational Research, 75*, 491–530.

Lemke, J. (1998). Multiplying meaning: Visual and verbal semiotics in scientific text. In J. R. Martin & R. Veel (Eds.), *Reading science: Critical and functional perspectives on discourses of science* (pp. 87–114). London, UK: Routledge.

Lopez, F., & McEneaney, E. (2012). State implementation of language acquisition policies and reading achievement among Hispanic students. *Educational Policy, 26*, 418–646.

Meara, P. (1980). Vocabulary acquisition: A neglected aspect of language learning. *Language Teaching and Linguistics: Abstracts, 13*(4), 221–246.

Michaels, S., Shouse, A. W., & Schweingruber, H. A. (2008). *Ready, set, SCIENCE!: Putting research to work in K-8 science classrooms.* Washington, D.C.: The National Academies Press.

National Center for Education Statistics. (2015). *English language learners in public schools.* U.S. Department of Education and the Institute of Education Sciences. Washington, D.C. In *The condition of education 2015 (NCES 2015–144).* Retrieved from https://nces.ed.gov/programs/coe/indicator_cgf.asp.

National Governors Association Center for Best Practices & Council of Chief State School Officers. (2010). *Common core state standards for English language arts and literacy in history/social studies, science, and technical subjects.* Washington, DC: Authors.

National Research Council. (2012). *A framework for K-12 science education: Practices, crosscutting concepts, and core ideas.* Washington, DC: National Academies Press.

NGSS Lead States.. (2013). *Next generation science standards: For states, by states*. Washington, DC: The National Academies Press. Retrieved from http://www.nextgenscience.org/sites/ngss/files/NGSS%20DCI%20Combined%2011.6.1.pdf.

Pearson, D., Hiebert, E. H., & Kamil, M. L. (2007). Theory and research into practice: A vocabulary assessment: What we know and what we need to learn. *Reading Research Quarterly*, *42*(2), 282–296.

Rapp, D. (2008). End of textbooks? What's stopping districts from ditching paper textbooks for good? Bureaucracy, budget woes, and inflexible teaching methods. In *Scholastic Magazine*. Retrieved from http://www.scholastic.com/browse/article.jsp?id=3750551.

Shamsudin, N. (2009). *Science text reading strategies: Learner's module*. Paper presented at the 2nd International Conference of Teaching and Learning, Universiti Teknologi MARA, Malaysia.

Valdés, G., Kibler, A., & Walqui, A. (2014). *Changes in the expertise of ESL professionals: Knowledge and action in an era of new standards*. Alexandria, VA: TESOL.

Samina Hadi-Tabassum is Assistant Professor in the Department of Curriculum and Instruction at Northern Illinois University (USA). Her research focuses on dual language education, science education, and multicultural education. Her first book publication, *Language, Space and Power: A Critical Look at Bilingual Education*, is an award-winning book capturing the ethnographic life of a dual language classroom. Her work has appeared in *Educational Leadership, Phi Delta Kappan, and Education Week.*

Emily Reardon is an Adjunct Professor in the School of Education at Loyola University in Chicago, IL (USA). She teaches courses in Curriculum & Instruction, Middle School Methodology, Assessment for ELs, and Theories and Foundations of Bilingual Education. She is in her eleventh year as a middle school science teacher to predominantly English Learners in the Chicago Public Schools. Additionally, she leads professional development for Chicago teachers focused on the Next Generation Science Standards and specifically the SEPUP (Science Education for Public Understanding Program) Life Science curriculum out of University of California - Berkeley.

4

Preparing Science Teachers for Project-Based, Integrated, Collaborative Instruction

Carrie L. McDermott and Andrea Honigsfeld

Students in K-12 classrooms must be actively engaged in all three dimensions of the science framework (Scientific and Engineering Practices, Cross-Cutting Concepts, and Disciplinary Core Ideas) to promote both language and science learning. (See Table 4.1 for select elements of this framework). Under the Next Generation Science Standards (NGSS) (NGSS Lead States 2013), science education is expected to reflect how science functions in the real world.

The framework is designed to help realize a vision for education in the sciences and engineering in which students, over multiple years of school, actively engage in scientific and engineering practices and apply cross-cutting concepts to deepen their understanding of the core ideas in these fields (National Academy of Sciences 2012, p. 10). English language learners (ELLs) face the dual challenge of mastering difficult core content and academic language. The rigor of learning experiences has increased across all disciplines, and ELLs are expected to effectively

C.L. McDermott (✉) · A. Honigsfeld
Division of Education, Molloy College, Rockville Centre, NY, USA
e-mail: cmcdermott@molloy.edu; ahonigsfeld@molloy.edu

© The Author(s) 2017

L.C. de Oliveira, K. Campbell Wilcox (eds.), *Teaching Science to English Language Learners*, DOI 10.1007/978-3-319-53594-4_4

Table 4.1 Three dimensions of the science framework

Scientific and engineering practices	Crosscutting concepts	Select disciplinary core ideas
1. Asking questions (for science) and defining problems (for engineering)	1. Patterns, similarity, and diversity	Physical Sciences
2. Developing and using models	2. Cause and effect: Mechanism and explanation	PS 1: Matter and its interactions
3. Planning and carrying out investigations	3. Scale, proportion, and quantity	PS 2: Motion and stability: Forces and interactions
4. Analyzing and interpreting data	4. Systems and system models	PS 3: Energy
5. Using mathematics and computational thinking	5. Energy and matter: Flows, cycles, and conservation	PS 4: Waves and their applications in technologies for information transfer
6. Constructing explanations (for science) and designing solutions (for engineering)	6. Structure and function	Engineering, Technology, and the Applications of Science
7. Engaging in argument from evidence	7. Stability and change	ETS 1: Engineering design
8. Obtaining, evaluating, and communicating information		ETS 2: Links among engineering, technology, science, and society

present ideas and comprehend information in academic English with a level of fluency necessary to be successful in the U.S. P-12 schools, with the goal for success after graduation in the workplace and/or college. In response to this growing need, there has been a call to shift the process of learning, especially in science, through the development of NGSS. This chapter is organized around unpacking three key approaches by defining each, offering research support for them and suggesting key steps to help prepare science teachers to effectively implement them. The three approaches are (a) project-based, inquiry-based science instruction, (b) integration of content and language instruction, and (c) teacher collaboration and co-teaching for ELLs in core science courses.

Scientific and engineering practices are the processes that students use to make sense of the real world coupled with the useful and applicable knowledge they gain. Pre-service science teachers need to be aware that ELLs grapple with science through planning and conducting investigations; creating and using models; constructing explanations and designing solutions; and obtaining, evaluating, and communicating information related to the cross-cutting concepts and disciplinary core ideas of science.

Cross-cutting concepts assist in understanding how the various components of subject matter, in this case science, interconnect through an organizational schema to help students relate to, apply, and communicate overall knowledge of a particular subject matter. Disciplinary core ideas are the specific branches of science studied in the classroom. These include earth, space science, physical, and life sciences in addition to engineering, technology, and the application of science; there are significant differences between students busy with work and students actively engaged in the learning process. This engagement allows students to navigate learning and become involved in concepts or ideas they find intellectually stimulating through the guidance of the teacher. It is part of the educator's responsibility to see equally to two things: First, that the problem grows out of the conditions of the experience being had in the present, and that it is within the range of the capacity of students; and, secondly, that it is such that it arouses in the learner an active quest for information and for production of new ideas (Dewey 1938/1963, p. 79).

NGSS continues to move science education away from students learning through memorization of facts (Alberts 2000) to a deeper understanding of knowledge-building by being a scientist in the classroom and learning to work as a team of scientists to discover meaning-making in the world around them.

Academic skills are permeated by language across all disciplines and play an essential role in the dynamics of language acquisition. "All students face language and literacy challenges and opportunities that are specific to science; such challenges and opportunities are amplified for ELLs and for other English speakers with limited standard English language and literacy development" (Lee et al. 2013 pp. 222–223). Teachers must be prepared to act as facilitators for their students' language and content development. Thus, we suggest a three-pronged approach to prepare science teachers for implementing the new standards with linguistically diverse student populations. This consists of (a) project-based, inquiry-based science instruction, (b) integration of content and language, and (c) a collaborative instructional service delivery including co-teaching core science courses.

Project-Based, Inquiry-Based Science Instruction

"Learning by doing... doing with understanding" (Barron et al. 1998, pp. 272, 276) and using tasks that are authentic to science (Lehrer and Schauble 2006) through project-based approaches are a way to make learning deeper and more applicable for students in relation to the world. This type of student-centered learning is aimed at building knowledge through inquiry both across and within content areas, over an extended period of time. This alters the teacher's role from a content deliverer to a scaffolder focusing on creating circumstances for students to learn as they engage in language learning through scientific activity.

What Is Project-Based, Inquiry-Based Science Instruction?

Project-based, inquiry-focused science is centered on practice and application. Inquiry instruction places higher demands on students in terms of participation, personal responsibility for learning, and intellectual effort (Blumenfeld et al. 2006). Collaborative teams of students become the scientists who examine the world by engaging in the processes of practice by using the tools necessary to meet the challenges of the world around them. In addition to learning the facts, concepts are brought together with new information and organized in a way that allows for a deeper understanding of natural phenomena (Harris and Rooks 2010) through language-based approaches to content. This allows students to have opportunities to apply ideas, use them to explain and predict phenomena, and make connections between scientific practices, cross-cutting concepts, and disciplinary core ideas.

Motion and stability: Forces and interactions are one of the core ideas for Physical Science within the three dimensions of the Science

Framework (as referred to earlier in this chapter). Motion and stability are directly impacted by gravity, friction, balanced and unbalances forces, inertia, centripetal force, and momentum. Newton changed how we understand these concepts by describing how movement occurs in the Universe with his three Laws of Motion. The first two laws refer to the forces acting on one object, focusing on the relationship between unbalanced forces and the motion associated with it. The third law focuses on forces acting on two different objects. Following are two examples of project-based science activities related to this topic, such as the popular Egg Drop Project and the Bridge Building Project. For each task within the project, it is important for the pre-service teacher to identify (1) the overall key concepts, (2) the language functions necessary to help students succeed, (3) the important content-relevant elements students will need to be successful within each task and goal, and (4) the guiding points, framework, or scaffolding that will help students complete this project.

Egg Drop Project

- *Task1*: Students work in teams of two to hypothesize, problem solve, and critically think about how to safely land a non-human (raw egg) on Earth.
  - *Student Goal*—Students construct a written plan including their hypothesis, problem, and at least two solutions with a rationale.
  - *Task 2*: Students construct a mechanism for the non-human (raw egg) to safely land on Earth for a trial experiment.
  - *Student Goal*: Students build something, made only from household products, to house the raw egg and protect it from breaking in a fall from a second or third story.
  - *Task 3*: Students complete the trial experiment.
  - *Student Goal*: (1) Students use their mechanism in the trial experiment and (2) problem solve through discussions and research to make revisions to all components from Task 1.

- *Task 4*: Students rebuild and complete the final experiment (videotaped).
- *Student Goal*: (1) Students revise their mechanism for the final experiment, (2) create a rationale for the amendments, and (3) conduct the final experiment.
- *Task 5*: Students watch the video and reflect on the designs.
- Student Goal: Students reflect on their peers' designs and write a description of what the model prototype should be based on their findings.

Bridge Research, Design, and Building

- *Task 1:* Students work in small *Research Teams* (three–four students) to explore various types of bridge structures found throughout the world. They use this information to investigate the purpose, strength, optimal conditions, and longevity of each design.
- *Student Goal:* Create a graphic organizer to indicate findings. This must include the type, purpose, an image, important facts, best locations, and any additional pertinent information.
- *Task 2:* Teams work together to learn about various aspects of an open case related to a bridge collapse. Each *Research Team* will have different information pertaining to the same case. Teams will work together to gather as much data as possible.
- *Student Goals:* (1) Work together to ask questions and define problems related to the case study, (2) construct an explanation of facts identifying the scenario and possible resolutions, (3) jigsawing with other teams to find out more information about the case, and (4) working in a fishbowl scenario to discuss the findings and solve the case.
- *Task 3:* A local charity is in need of a design team to help construct a bridge to help people in a remote area travel for supplies and medical treatment. The charity would like to know more about the area, what structures would be most beneficial, what problems will be encountered, how these problems could be alleviated, cost

factor, and sustainability. A scale model must accompany the proposal.

- *Student Goal:* (1) Construct a proposal for the charity highlighting each of the points requested along with additional information, (2) design and sketch a bridge, (3) construct a scale model, and (4) create a presentation to sell this proposal to the charity (note: the presentation will be in class with community members, teachers, and administrators).

How to Prepare Pre-Service Science Teachers for Success with Project-Based, Inquiry-Based Science Instruction

Effective project-based and student-centered learning environments only lead to success if the teacher offers skilled and thoughtful guidance (Harris and Rooks 2010). The classroom environment is predicated on engaging student learning through inquiry, active student involvement, and practice (National Research Council [NRC] 2000). Effective cooperative learning techniques must be an integral part of the overall classroom environment as students learn to inquire, reflect, and assess their learning.

Teachers learn to gauge the gap between what students are capable of doing and what knowledge they need to complete a task. Project-based learning offers teachers a way to scaffold information to close this gap and help students find success in the classroom. Hart et al. (2000) also suggested that providing the reason for the task(s) a student is working on adds purpose and gives students rationale and relevance for what they are doing, making it a worthwhile activity. These are the reasons it is important for in-service teachers to identify the concepts, language functions, relevant content, and framework to make project-based learning successful for ELLs in the science classroom. Another point to keep in mind is that rigor and complexity of tasks may increase the likelihood of a shift from "doing with understanding... [to only] following the procedures" (Barron et al. 1998, p. 276).

Research Support for Project-Based Learning

John Dewey has guided the implementation of project-based learning since the early 1900s. He stated, "We do not learn from experience . . . we learn from reflecting on experience" (p. 78). Students of the twenty-first Century are expected to solve world problems, think critically, innovate, effectively communicate, and much more. Teachers are charged with embracing all of these and harnessing the power of the interactive mind. Project-based learning helps students acquire knowledge, practice skills, and take risks in learning while preparing them for *deeper thinking* (Boss 2015) leading them to make informed decisions about the world around them. According to Larmer et al. (2015), the *gold-standard* of project-based learning include three main components: learning goals, elements of project design, and teaching practices. These standards used in conjunction with teaching content through language as discussed in de Oliveira's (2016) research on language-based approaches to content instruction (LACI) in the classroom where teachers intentionally use the target language to introduce ideas, build relationships, and help students construct a comprehensive understanding of the material.

Research has shown that project-based learning benefits student achievement. Higher scores were reported on state assessments focusing on science content knowledge and processing skills for ELLs, African-Americans, and Hispanic students in an urban middle school who were taught through project-based learning initiatives (Geier et al. 2008). Education for Sustainable Communities (2011) reported that schools implementing project-based learning showed progress in closing the achievement gap for ELLs, minority students, and those on free and reduced lunch programs (Boss 2015). Overall, these researchers suggest that teachers have found that leveling the playing field through project-based learning to be beneficial to all students. In fact this is further corroborated by the idea of using open-ended problem solving curricula; middle and high school teachers found a reduction in inequalities related to language, culture, and ethnicity in their schools (Boaler 2002) (Table 4.2).

Table 4.2 The characteristics of effective science lesson delivery for ELLs

Quality of lesson design for ELLs	Quality of non-interactive/dialogic (NI/D) for ELLs
• Resources available contribute to accomplishing the purpose of content knowledge, language instruction, and language acquisition. • Reflects careful planning and organization. • Strategies and activities reflect attention to students' preparedness, prior experience, and language proficiency. • Strategies and activities reflect attention to issues of access, equity, and diversity. • Incorporates tasks, roles, and interactions consistent with investigative science. • Encourages collaboration among students. • Provides adequate time, strategies, and structure for sense-making and reflection to meet diverse students needs and language proficiency levels.	• Teacher appears confident in ability to teach science through language. • Teacher's classroom management enhances quality of lesson. • Pace is appropriate for developmental level/needs and language proficiency levels of students. • Teacher is able to adjust instruction according to level of students' understanding, ability, and language proficiency level. • Instructional strategies are consistent with investigative science and language acquisition. • Teacher's questioning enhances development of students' understanding/problem solving and language skills.
Quality of integrated science content for ELLs	**Quality of classroom culture for ELLs**
• Content is significant, worthwhile, and taught through language. • Content information is accurate and builds understanding and acquisition of English. • Content is appropriate for developmental levels and proficiency levels of students. • Teacher displays understanding of concepts in both science content and language acquisition.	• Climate of respect for students' ideas, questions, contributions, and linguistic/cultural diversity is evident. • Active participation of all is encouraged and valued. • Interactions reflect working relationship between teachers and students. • Climate encourages students to generate ideas and questions that are both content and language relevant.

(*continued*)

Table 4.2 (continued)

Quality of lesson design for ELLs	Quality of non-interactive/dialogic (NI/D) for ELLs
• Students are intellectually and linguistically engaged with important ideas. • Appropriate cross-cutting connections are made and are appropriate for proficiency levels of students. • Subject is portrayed as dynamic body of knowledge with significant focus on language development. • Degree of sense making is appropriate for this lesson, diverse needs of students, and language proficiency levels.	• Intellectual rigor, constructive criticism, and challenging of ideas and language ability are evident. • A collaborative culture is evident, and students are challenged to take risks in learning new and difficult information while strengthening their language acquisition.

Adapted from Weiss et al. 2003; Tweed 2009

Integration of Science Content and Language

There is a direct connection between children's cognitive development and language acquisition. As children acquire new information and knowledge, they need support when learning new skills to accommodate language learning (Schleppegrell 2012). Student learning in content areas becomes more cognitively demanding over time. Educators need to spend time teaching students the academic language necessary to be successful in school. Academic language has been defined in a range of different ways by researchers and practitioners, based on their focus or philosophy. For example, Gottlieb and Ernst-Slavit (2014) noted, "academic language refers to the language used in school to acquire new or deeper understanding of the content and to communicate that understanding to others" (p. 2). Thus academic language as the preferred style of communication in schools and the vehicle for content attainment must be recognized and directly taught in all content areas including science. One pathway to this approach to teaching is the integration of content and language instruction.

What Is Integration of Content and Language

When content and language are systematically integrated within the context of a single secondary science content course, the goal of instruction goes beyond mastery of academic content. The course is designed to support students' mastery of the language and literacy skills associated with the content area as well, thus preparing students to speak and write (not just read and listen) the language of scientific inquiry. Baker et al. (2014) emphasized the importance of incorporating explicit focus on academic language instruction in the content areas.

Theoretical and Research Support for Integration of Content and Language

Integrating content and language requires a complex skill set that educators are rarely prepared in during their pre-service years nor supported with during in-service professional learning programs. Turkan et al. (2014) proposed a conceptual framework referred to as *disciplinary linguistic knowledge* (DLK). This type of knowledge is unique to each content area or discipline and is instrumental in both identifying the language and literacy demands of a particular content area (such as science), and in helping ELLs develop receptive and productive language skills associated with the target content. More specifically, DLK refers to "(a) the ability to identify the linguistic features and choices that are appropriate to the disciplinary discourse and (b) the ability to model these for students," as well as "teachers' knowledge of a particular disciplinary discourse and involves knowledge for (a) identifying linguistic features of the disciplinary discourse and (b) modeling for ELLs how to communicate meaning in the discipline and engaging them in using the language of the discipline orally or in writing" (p. 9). The goal of helping pre-service teachers develop DLK is to ensure that their students will have access to the complex language and content of the discipline as well as be given the opportunity to actively participate in science classes.

How to Prepare Science Teachers for Success with Content and Language Integration

To be most supportive of English learners' language and disciplinary literacy development, science educators need a practical, accessible way to think about the academic language needed for success in the science classroom. Members of the Complex Language Development Network also developed a framework for academic language and literacy development that consists of three dimensions of academic language depicted in Table 4.3 (O'Hara et al. 2013). We adapted the framework by adding specific challenges faced by ELLs in the science classroom and expanding the framework to include key instructional strategies that helps connect theory to practice while preparing science teachers for working with ELLs.

Collaborative Instructional Service Delivery

Focused, shared collaborative practices among teachers are undertaken time and again because of a compelling urgency to accommodate and differentiate instruction for the sake of all students. Without a doubt, capacity building should be centered on developing ways to support diverse learners in general education classes such as building students' background knowledge, using flexible groupings, making lessons relevant to students' lived experiences, and engaging all learners in complex, critical-thinking tasks (Honigsfeld and Dove 2010). Yet, the question remains how science teachers and ESOL specialists can most effectively co-deliver instruction that is both meaningful and rigorous for all.

What Is Collaborative Instructional Service Delivery?

Rather than subjecting ELLs to a disjointed, fragmented school day, in which they move from class to class where teachers each deliver their own core content with little or no attention to the academic language of discipline, we suggest a more cohesive, collaborative approach to

Table 4.3 Academic language dimensions, features, challenges for ELLs and key instructional practices

Dimension	Academic language features	Challenges for ELLs in the science classroom	Key instructional practices for science teachers
Word-level (Vocabulary or phrases)	Generic and content-specific academic terms; Figurative and idiomatic expressions; Words with multiple meanings Roots and affixes	• Volume of technical vocabulary needed • Nuances of word meanings • Scientific phrases and expressions	• Exposure to science vocabulary through work with varied tasks • An interactive environment in which student-to-student exchanges are encouraged • Word learning strategies
Sentence-level	Sentence structure Sentence length Grammatical structures Pronouns Context clues Proverbs	• Complex sentences • Advanced grammatical features (passive voice)	• Sentence deconstruction (sentence chunking with discussion on the form and meaning of each segment) • Sentence starters and frames
Text-level	Text organization Text types and genres Text structure Text density Clarity and coherence	• Reading and lexile levels • Complexity of ideas • Background knowledge students need to comprehend science content • Styles and structures unique to scientific texts	• Read alouds and shared reading • Scaffolded independent reading • Comprehension strategy instruction within science • Inquiry groups that work with print and non-print text • Text analysis • Text annotation

Adapted from O'Hara et al. (2013).

instruction. It has been noted that "the long-standing culture of teacher isolation and individualism, together with teachers' preference to preserve their individual autonomy, may hinder deep-level collaboration to occur" (Vangrieken et al. 2015, p. 36). English learners can no longer be isolated and removed from the classroom for extended periods of time (US DOE 2015), instead there is a strong call for including ELLs in content area with carefully planned and delivered instruction. In a collaborative, integrated, model of instruction, teachers engage in a range of formal collaborative practices to support ELLs' linguistic and academic development (Honigsfeld and Dove 2010, 2014). These activities include (1) joint planning, (2) curriculum mapping and alignment, (3) parallel teaching, (4) co-developing instructional materials, (5) collaborative assessment of student work, (6) co-teaching, (7) joint professional development, and (8) teacher research (See Table 4.4 for a summary of these practices and their implication for science teacher education).

Theoretical and Research Support for Collaboration and Co-teaching

Teacher collaboration and an integrated, collaborative service delivery option for the sake of ELLs is an emerging topic of interest in the TESOL field. To further our discussion on how to integrate collaboration and co-teaching into science teacher education, we base our discussion on select theories and evidence-based practices, including inclusive pedagogy and co-teaching (Honigsfeld and Dove 2010).

Inclusive pedagogy offers a major theoretical framework and evidence-based practice that is helpful in understanding what science teacher education for ELLs should look like. It is based on the premise that teachers must make learning accessible to all. Spratt and Florian (2013) suggested focusing on each child's potential to learn and not on their deficiencies. They stated that, "human diversity is seen within the model of inclusive pedagogy as a strength, rather than a problem, as children work together, sharing ideas and learning from their interactions with each other" (p. 135). While inclusive pedagogy is often discussed in the Pre-Kindergarten-12 context, it provides an important

Table 4.4 Collaborative practices and their implications for science teacher education

Collaborative practices aligned to instruction	Goals	Implications for science teacher education
Joint planning	• To establish attainable yet rigorous learning targets • To share instructional routines and strategies • To align instructional content • To design appropriate formative and summative assessment measures	• Teachers must learn to prepare daily lesson plans and unit plans reflective of the following: • language and since content objectives • knowledge of diverse ELLs' needs • strategically selected instructional accommodations and accelerations • differentiated instruction according to students' academic and linguistic abilities
Curriculum mapping and alignment	• To plan and align instruction for a longer period of time • To have an overall guide for joint planning, parallel teaching, and co-instruction	• Teachers must learn to infuse rigor, relevance, and research-informed approaches into the science curriculum • Teachers must commit to instructional intensity in the planned and taught science curriculum for ELLs
Parallel teaching	• To accelerate ELLs' knowledge and understanding of mainstream curricula • To ensure that what happens during ESL lessons parallels general class instruction	• Teachers must be prepared for coordinating and sharing lesson goals and objectives with general education colleagues • Teachers must learn to pre-teach or re-teach essential concepts and related language and disciplinary literacy skills.
Co-developing instructional materials	• To scaffold instructional materials	• Teachers must learn to develop differentiated, tiered, multi-level

(*continued*)

Table 4.4 (continued)

Collaborative practices aligned to instruction	Goals	Implications for science teacher education
	• To select essential materials that support accelerated learning	instructional resources as well as divide complex materials or tasks into manageable segments while also helping ELLs master essential learning strategies
Collaborative assessment of student work	To Jointly examine ELLs' language and academic performance Analyze student data and identify areas that need improvement or targeted intervention	• Teachers must learn to develop and administer formative and summative assessment measures, as well as evaluate the student outcomes gained from them • Teachers must learn to set language and content goals for ELLs and use assessment data collaboratively
Co-teaching	Co-deliver instruction through differentiated instruction Use various models of instruction to establish equity between co-teaching partners	• Teachers must learn to establish co-equal partnerships with other teachers and share ownership of teaching ELLs • Teachers must learn to engage in the entire collaborative instructional cycle with their colleagues
Joint professional learning	Enhance pedagogical knowledge, skills, and dispositions about ELLs Establish a shared understanding about ELLs' needs, best practices, and effective strategies Explore new and emerging directions in ELD/ESL education	• Teachers must have sustained opportunities and commitment to engage in • Learning with colleagues • Applying their new learning to teaching • Reflecting on their new learning • Showcasing their practices while inviting feedback from critical friends

Table 4.4 (continued)

Collaborative practices aligned to instruction	Goals	Implications for science teacher education
Collaborative teacher research	Collect and analyze student data in response to instructional practice Document differential student progress in response to interventions	• Teachers must understand the use and meaning of educational research • Teachers must learn to design and implement classroom-based research or participatory action research

Adapted from Honigsfeld and Dove (2014).

theoretical framework for science teacher education as well. Successful inclusive pedagogy heavily relies on teacher collaboration, often including or centering around co-teaching practices that allow two or more educators to plan, deliver, and assess instruction for the sake of special populations while also setting challenging educational goals and delivering differentiated instruction for all students.

While a considerable volume of research has focused on collaboration among general- and special-education teachers (see Scruggs et al. 2007 for a metasynthesis), similar attention to ESL collaboration is now emerging (see, for example, Honigsfeld and Dove 2012). Several co-teaching frameworks have been utilized within the context of special education as well as ESOL inclusion (Villa et al., 2013; Vaughn et al. 1997; Murawski 2009). Among others, Davison (2006) extensively researched collaboration among ESL and content-area teachers with a special emphasis on the nature and challenges of developing collaborative and co-teaching relationships. She used the term *partnership teaching* and emphasized that "it builds on the concept of co-operative teaching by linking the work of two teachers, or indeed a whole department/year team or other partners, with plans for curriculum development and staff development across the school" (p. 455).

How to Prepare Science Teachers for Success with Collaboration and Co-teaching

We suggest that collaboration and co-teaching practices should be better addressed in programs for pre-service and in-service science teachers. A recent study noted that teachers across the U.S. spend only about 3% of their teaching day collaborating with colleagues (Bill and Melinda Gates Foundation 2012), whereas in other countries teacher collaboration is viewed as a key factor in instructional improvement (Mirel and Golding 2012).

Pre-service and in-service science teachers who work with ELLs need to

- engage in productive collaborative conversations that help them refine their professional skill sets regarding ELLs' needs;
- have frequent, structured opportunities for exploring the NGSS and necessary adaptations and modifications for ELLs;
- review curriculum materials (textbooks, school/district/state curricular frameworks, scope and sequence charts and teacher created instructional materials);
- have common preparation time built into teachers' schedules paired with accountability for their time used productively (use of agendas and protocols are strongly urged);
- understand their own strengths, areas that need growth or support, and varying teaching styles;
- engage in a systematic use of formative and summative assessment tools that inform their instruction;
- understand, regularly review, and analyze student assessment data, which, in turn, will inform instruction;
- participate in ongoing professional learning opportunities that focus on improving their own practice and increasing student learning;
- discuss student performance data and generate possible explanations for patterns of uneven student outcomes; and
- plan, coordinate, implement, and assess the outcomes of appropriate interventions for students in need of additional academic or linguistic support.

Conclusion

Integrating concepts into project-based learning ensures students gain knowledge through doing (Dewey, 1933). Educators committed to PBL promote student investigation and critical thinking skills, which are essential to the success of students learning rigorous science content while acquiring language. The further pursuit of helping students link knowledge to the real world also fosters a passion for life-long learning which creates a deeper understanding of how things work, and identifies how each individual impacts the world through science, engineering, inquiry, invention, and technological advancement.

Future science teachers need (a) project-based, inquiry-based science instruction, (b) integration of content and language, and (c) a collaborative instructional service delivery including co-teaching core science courses. The goals of these elements are to ensure students are actively engaged in all three dimensions of the science framework as a means to master academic language and science learning through fluency in practice and application. The application and practice prepare students for the world they know as well as the one they will be masterminds in creating. As educators, we are preparing students to see what is in front of them as well as the world beyond their greatest expectations.

References

Alberts, B. (2000). Some thoughts of a scientist on inquiry. In J. Minstrell & E. Van Zee (Eds.), *Inquiring into inquiry learning and teaching in science* (pp. 3–13). Washington, DC: American Association for the Advancement of Science.

Baker, S., Lesaux, N., Jayanthi, M., Dimino, J., Proctor, C. P., Morris, J., Gersten, R., Haymond, K., Kieffer, M. J., Linan-Thompson, S., & Newman-Gonchar, R. (2014). *Teaching academic content and literacy to English learners in elementary and middle school.* IES Practice Guide. NCEE 2014–4012. Retrieved from http://ies.ed.gov/ncee/wwc/pdf/practice_guides/english_learners_pg_040114.pdf.

Barron, B., Schwartz, D., Vye, N., Moore, A., Petrosino, A., Zech, L., Bransford, J., & Cognition, T. and Technology Group at Vanderbilt. (1998). Doing with understanding: Lessons from research on problem and project-based learning. *The Journal of the Learning Sciences, 7*, 271–311. Retrieved from http://web.mit.edu/monicaru/Public/old%20stuff/For%20Dava/Grad%20Library.Data/PDF/Brigid_1998DoingwithUnderstanding-LessonsfromResearchonProblem-andProject-BasedLearning-1896588801/Brigid_1998DoingwithUnderstanding-LessonsfromResearchonProblem-andProject-BasedLearning.pdf.

Bill & Melinda Gates Foundation. (2012). *Primary sources 2012: America's teachers on the teaching profession*. Seattle, WA: Author.

Blumenfeld, P. C., Kempler, T. M., & Krajcik, J. S. (2006). Motivation and cognitive engagement in learning environments. In R. K. Sawyer (Ed.), *Cambridge handbook of the learning sciences* (pp. 475–488). New York, NY: Cambridge University Press.

Boaler, J. (2002). Learning from teaching: exploring the relationship between reform curriculum and equity. *Journal for Research in Mathematics Education, 33*(4), 239–258. Retrieved from https://pdfs.semanticscholar.org/6292/668fbace9d59357117f05053a9aa7b04e8b9.pdf.

Boss, S. (2015). *Implementing project-based learning*. Bloomington, IN: Solution Tree Press.

Davison, C. (2006). Collaboration between ESL and content area teachers: How do we know when we are doing it right?. *The International Journal of Bilingual Education and Bilingualism, 9*(4), 454–475.

de Oliveira, L. C. (2016). A language-based approach to content instruction (LACI) for English language learners: Examples from two elementary teachers. *International Multilingual Research Journal, 10*(3), 217–231. doi: 10.1080/19313152.2016.118591.

Dewey, J. (1933). *How we think*. Boston, MA: D. C. Heath & Co..

Dewey, J. (1938/1963). *Experience and education*. New York, NY: Collier Books.

Education for Sustainable Communities. (2011). *Expeditionary learning*. Retrieved from https://sites.tufts.edu/uep284edu/national-happenings/expeditionary-learning/.

Geier, R., Blumenfeld, P. C., Marx, R. W., Krajcik, J. S., Fishman, B., Soloway, E., & Clay-Chambers, J. (2008), Standardized test outcomes for students engaged in inquiry-based science curricula in the context of urban reform. *Journal of Research in* S. Teaching *45*, 922–939. doi: 10.1002/tea.20248.

Gottlieb, M., & Ernst-Slavit, G. (2014). *Academic language in diverse classrooms: Definitions and contexts*. Thousand Oaks, CA: Corwin.

Harris, C. J., & Rooks, D. L. (2010). Managing inquiry-based science: Challenges in enacting complex science instruction in elementary and middle school classrooms. *Journal of Science Teacher Education, 21*, 227–240. doi: 10.1007/s10972-009-9172-5.

Hart, C., Mulhall, P., Berry, A., Loughran, J., & Gunstone, R. (2000). What is the purpose of this experiment? Or can students learn something from doing experiments?. *Journal of Research in Science Teaching, 37*(7), 655–675.

Honigsfeld, A., & Dove, M. G. (2010). *Collaboration and co-teaching: Strategies for English learners*. Thousand Oaks, CA: Corwin.

Honigsfeld, A., & Dove, M. G. (Eds.) (2012). *Coteaching and other collaborative practices in the EFL/ESL classroom: Rationale, research, reflections, and recommendations*. Charlotte, NC: Information Age Publishing.

Honigsfeld, A., & Dove, M. G. (2014). *Collaboration and co-teaching for English learners: A leader's guide*. Thousand Oaks, CA: Corwin.

Larmer, J., Mergendoller, J., & Boss, S. (2015). *Setting the standard for project based learning*. Alexandria, VA: ASCD.

Lee, O., Quinn, H., & Valdes, G. (2013). Science and language for English language learners in relation to next generation science standards and with implications for common core state standards for English language arts and mathematics. *Educational Researcher, 42*(4), 223–233. doi: 10.3102/0013189x13480524.

Lehrer, R., & Schauble, L. (2006). *Cultivating model-based reasoning in science education*. In R. Keith Sawyer (Ed.), *Cambridge handbook of the learning sciences* (pp. 371–387). Cambridge, MA: Cambridge University Press.

Mirel, J., & Goldin, S. (2012). *Alone in the classroom: Why teachers are too isolated*. Retrieved from *http://www.theatlantic.com/national/archive/2012/04/alone-in-the-classroom-why-teachers-are-too-isolated/255976/*.

Murawski, W. W. (2009). *Collaborative teaching in elementary schools: Making the co-teaching marriage work!*. Thousand Oaks, CA: Corwin.

National Academy of Sciences. (2012). *A framework for K-12 science education: Practices, crosscutting concepts, and core ideas*. Washington, DC: The National Academies Press.

National Research Council. (2000). *Inquiry and the national science education standards*. Washington, DC: National Academy Press.

NGSS Lead States. (2013). *Next generation science standards: For states, by states*. Washington, DC: The National Academies Press. Retrieved from https://www.nap.edu/read/18290

O'Hara, S., Zwiers, J., & Pritchard, R. (2013). *Framing the development of complex language and literacy*. Retrieved from http://www.aldnetwork.org/sites/default/files/pictures/aldn_brief_2013.pdf.

Schleppegrell, M. J. (2012). Academic language in teaching and learning: introduction to the special issue. *The Elementary School Journal, 112*(3), 409–418.

Scruggs, T. E., Mastropieri, M. A., & McDuffie, K. A. (2007). Co-teaching in inclusive classrooms: A metasynthesis of qualitative research. *Exceptional Children, 73*, 392–416.

Spratt, J., & Florian, L. (2013). Applying the principles of inclusive pedagogy in initial teacher education: From university based course to classroom action. *Revista de InvestigaciÓn en EducaciÓn, 11*(3), 133–140.

Turkan, S., de Oliveira, L. C., Lee, O., & Phelps, G. (2014). Proposing the knowledge base for teaching academic content to English language learners: Disciplinary linguistic knowledge. *Teachers College Record, 116*(3), 1–30.

Tweed, A. (2009). *Designing effective science instruction*. Arlington, VA: Mid-continent Research for Education and Learning (McREL.

U.S. Department of Education (DOE). (2015). *English learner toolkit for state and local education agencies (SEAs and LEAs)*. Retrieved from http://www2.ed.gov/about/offices/list/oela/english-learner-toolkit/index.html.

Vangrieken, K., Dochy, F., Raes, E., & Kyndt, E. (2015). Teacher collaboration: A systematic review. *Educational Research Review, 15*, 17–40.

Vaughn, S., Schumm, J. S., & Arguelles, M. E. (1997). The ABCDEs of co-teaching. *Teaching Exceptional Children, 30*(2), 4–10.

Villa, R. A., Thousand, J. S., & Nevin, A. I. (2013). *A guide to co-teaching: New lessons and strategies to facilitate student learning* (3rd ed.) Thousand Oaks, CA: Corwin.

Weiss, I. R., Pasley, J. D., Smith, P. S., Banilower, E., & Heck, D. (2003). *Looking inside the classroom: A study of K-12 mathematics and science education in the United States*. Chapel Hill, NC: Horizon Research.

Carrie L. McDermott is Assistant Professor in the Division of Education at Molloy College, Rockville Centre, New York with concentrations in action research, cultural and linguistic diversity, ESOL methodology, theory, and acquisition. Additionally, she trains/coaches teachers in integrated and collaborative instruction and works with administrators to supervise and evaluate these practices. She is currently involved in several research projects involving the evolution of integrated reading comprehension applications for English language learners, co-teaching practices, and graduate education program impact.

Andrea Honigsfeld is Associate Dean and Director of the Doctoral Program (Educational Leadership for Diverse Learning Communities) at Molloy College, Rockville Centre, New York. She co-authored and co-directed a 3-year NYSED-funded Clinically-Rich Intensive Teacher Institute grant that supports ESOL certification and Bilingual Extension in high-needs schools on Long Island. A Fulbright scholar and coauthor of 17 books (3 of which are Corwin Press national best sellers) and over 60 articles, she is a nationally recognized expert on collaboration and co-teaching for English language learners.

5

Maximizing Science Teachers' Pedagogical Potential for Teaching Science Literacy to ELLs: Findings from a Professional Development Program

Clara Lee Brown and Mehmet Aydeniz

> Good teaching is not an accident. While some teachers are more naturally gifted than others, all effective teaching is the result of study, reflection, practice, and hard work. A teacher can never know enough about how a student learns, what impedes the student's learning, and how the teacher's instruction can increase the student's learning. Professional development is the only means for teachers to gain such knowledge. Whether students are high, low, or average achievers, they will learn more if their teachers regularly engage in high-quality professional development (Mizell 2010, p. 18).

It is estimated that there are over 11 million English Language Learners (ELLs) in the United States who are characterized by their below-grade-level academic English (National Center for Education Statistics [NCES] 2012). These students struggle because they are in the process of acquiring a new language, which is different from their native language. ELLs thus face

C.L. Brown (✉) · M. Aydeniz
Department of Theory and Practice in Teacher Education, The University of Tennessee, Knoxville, USA
e-mail: cbrown26@utk.edu; maydeniz@utk.edu

© The Author(s) 2017

L.C. de Oliveira, K. Campbell Wilcox (eds.), *Teaching Science to English Language Learners*, DOI 10.1007/978-3-319-53594-4_5

tremendous academic challenges resulting in test scores significantly lower than those of their peers as indicated in the recent National Assessment of Educational Progress (NAEP) report (NCES 2012). This discrepancy between general education students' and ELLs' academic achievement is even more pronounced in science. ELLs' science achievement is significantly lower than those of their fully English-speaking white peers (NCES 2012). These results indicate that traditional instructional methods are neither adequate nor effective in addressing the learning needs of ELLs in science, highlighting the urgency of the matter. Given the increasing number of ELLs in the U.S., it is necessary to explore effective instructional strategies that could help ELLs maximize their potential and achieve higher in science. Science educators need to develop instructional strategies specifically designed for ELLs so that they, too, can access high-quality curriculum and effectively engage in learning.

In this chapter, we briefly explain the current educational landscape regarding content literacy under the Common Core State Standards (CCSS) (National Governors Association Center for Best Practices (NGA Center), Council for Chief State School Officers (CCSSO) 2010) and discuss how critical the comprehension of informational texts is to ELLs' ability to construct and evaluate scientific arguments and explanations as Next Generation Science Standards (NGSS) (National Research council 2012) recommend. We then discuss the specific learning challenges that ELLs encounter in science learning due to linguistic barriers. Finally, we illustrate the positive impact of a year-long professional development (PD) program based on four high school science teachers who reported they still implement strategies and activities they learned in the PD program.

The Urgent Need for Professional Development

The advent of CCSS in 2010 unprecedentedly (and ambitiously) declared that all students graduating from high school in the U.S. should be college and career ready. CCSS made an explicit pedagogical shift in the ways in which science be taught. That is, content delivery alone is no longer considered adequate if the goal is to prepare all students to be ready for college and careers. The CCSS emphasize that content area literacy must be a high priority for content teachers and, accordingly, a heavy emphasis was put on informational text reading in content areas (NGA Center, CCSSO 2010).

The NGSS, joining in and agreeing with the rationale behind the CCSS, explicitly emphasized improving all students' abilities to develop evidence-based scientific arguments while effectively communicating with peers and evaluating the scientific validity of the arguments to build scientific knowledge. In order for the students of science to perform what NGSS calls for (e.g., reasoning a hypothesis and establishing a theory), highly proficient reading skills are fundamental in order to comprehend complex informational texts with precise and detailed comprehension (Lee et al. 2014). For example, Common Core science literacy standards for the 9th grade state the following:

> *Cite specific textual evidence to support analysis of science and technical texts, attending to the precise details of explanations or descriptions.* (**CCSS.ELA-Literacy.RST.9–10.1**)

As the standard clearly indicates, just getting the gist of what they read would not be nearly enough if ELLs are to effectively and successfully engage in constructing scientific arguments or "evaluating competing ideas and methods based on their merits," as suggested in NGSS. For ELLs, they must first decode and comprehend the informational text itself even before engaging in an argument accompanied by textual evidence using discipline-specific technical terms. Demonstrations of such competencies are a tall order for them when they are struggling with English.

Under the new curricular guidelines of both NGSS and CCSS, ELLs' ability to comprehend densely written informational science texts is extremely critical. To bring them up to speed seems to be an essential step to take so that they, too, get ready for college and career. While the policy makers seemed to have marched on along with the ideas and decisions of NGSS and CCSS, teachers in the science classroom might feel they are left behind with little support from the system (Lee et al. 2014; Lee et al. 2013). A nation-wide teacher survey, conducted by Education Week with 500 teachers, revealed only 8% of the survey teachers indicated that they felt "well prepared" to teach CCSS to ELLs (Education Week Research Center 2013). This may explain why ELLs are rarely accommodated in class (Lee and Fradd 1998). It is mostly because teachers are not prepared to assist them. In Brown (2003), a third-grader poignantly recounted her learning experience in class with her teacher: "Mrs. Smith helps me a little, I mean a little, but not so much" (para. 25). What is apparent is that teachers like Mrs. Smith, who are as nice as they can be toward ELLs, just do not know how to accommodate them. Even a third grade ELL in this study was highly aware of the fact that her learning difficulties were not being mediated by her teacher. A story like this sheds light on teachers' lack of preparedness to adequately teach ELLs. When ELLs do not learn the content in science classrooms, for instance, due to incomprehensible content instruction given by their teacher, they cannot be expected to succeed in science learning. Since the Coleman Report (1966), a strong correlation existing between teacher quality and student achievement has been well documented, along with the impact of teachers' instructional performance on student achievement (Darling-Hammond 2000; Hanushek 1992; Hanushek et al. 1998). According to Darling-Hammond (1999), teacher-quality variables are the most robust predictors, accounting for 40 to 60% of total variance in student achievement.

Despite the fact that the NGSS maintain that science, technology, engineering, and mathematics (STEM) are vital for the nation's next generation of students to compete in the twenty-first century economy, the achievement gap in science widens for minority students, especially for language minorities (NCES 2012). Ongoing, sustained PD is one of the most effective interventions in strengthening the link between teacher quality and student achievement (Lee et al. 2004; Yoon et al. 2007). Darling-Hammond et al. (2010) found that sustained teacher PD improved student learning in general when examining the impact of teacher PD among high achieving nations including South Korea, Singapore, and Sweden. An analysis reviewing evidence from 1,300 studies to investigate the ways in which PD affects student achievement reported that an average of 49 PD hours throughout a semester or two raised student achievement by 21% points (Yoon et al. 2007). Indeed, one of the prominent features of successful PD is the ongoing nature of the program that allows the process to pan out (Ball and Cohen 1999). Teachers themselves need to believe that substantive PD advances their pedagogical and content knowledge and, at the same time, that it extends opportunities for them to apply more practical methods in the classrooms (Hill 2009; McLaughlin and Talbert 2001, 2006).

According to the National Research Council (NRC) (2001), the major source of teachers' pedagogical knowledge is actually the PD, and abundant evidence in the literature shows that teacher PD is extremely critical if not the make-or-break undertaking. Yet in reality, 85% of California's science teachers reported that they did not receive any PD for 3 consecutive years after budget cuts in 2011 (Dorph et al. 2011). Another nationwide survey revealed that just 12.5% of all teachers across the nation received only one-day PD on teaching ELLs in the past three years (NCES, 2002). Although research recommends 10% of district budgets be designated for teacher PD, while it is difficult to pinpoint, studies suggest that only 1% to 3% of the district budget is used for PD on a national average (Miles et al. 2004). Research has shown that high-quality PD can indeed transform teachers' instructional practices (Darling-Hammond 2004; Gallimore et al. 2009). Studies also indicate that consistent and continuing PD support for teachers is key for implementing reform-based, high-quality instruction (Walqui and Van Lier 2010). The majority of science teachers, however, have limited pedagogical knowledge and skills needed to help ELLs develop science-related literacy (Lee 2005). In addition, if equal footing in the classroom is the goal for ELLs, then increasing teachers' instructional capacity is a critical first step to take.

Science Is Way More Than Learning About Science

Some erroneously hold on to the idea that learning science is about memorizing factual information and signs and symbols; however, others suggest that learning science is more like learning a new culture (Rosebery et al. 1992). As culture is about learning of a new world to be discovered, science learning is similarly about discovering nature and the world beyond that of facts and figures. More precisely, it is especially about ways of thinking, knowing, developing and communicating evidence-based explanations, which can be distinctively different from other types of content learning such as language arts. It is fair to say, then, that discovering a new world with a new language would be doubly challenging. Science learning is usually characterized more by conceptual learning, with less recognition on the role of language. Even before CCSS and NGSS, many (Fang 2006; Fang and Wei 2010; Gee 2005; Lemke 1990) discussed the importance of science teachers modeling the language of science to demonstrate specific language use at the discourse level in order to enhance conceptual learning. Brown and Ryoo (2008) also highlighted the interdependency of conceptual understanding in science and the science-specific register, asserting that conceptual understanding is highly hinged upon meaning-making process facilitated by language use. Aptly, Turkan et al. (2014) argue for "Disciplinary Linguistic Knowledge" (p. 9) by content teachers that would allow them to identify and to model necessary linguistic features that aid ELLs to "access" content learning (p. 10). In other words, science teachers' domain-specific knowledge alone is not sufficient, if they are to effectively instruct science content to ELLs.

Some cite poor informational reading skills as the culprit behind poor science achievement (Duke 2004). For ELLs, this is particularly true, and there are a multitude of reasons why informational text is especially difficult for ELLs. As they need to develop competencies in acquisition and application of science concepts and processes, the language of science and its expository discourse present unique challenges for ELLs who learn through a yet-to-be-mastered language. Science concepts are often highly abstract and may be unfamiliar to them; as such, the language functions of science add more hurdles to the challenges. Zwiers (2007) posits that, to be successful in school, ELLs need "academic capital," applying the concepts of Bourdieu's (1977, 1986) linguistic and cultural capitals. Zwiers' notion of academic capital refers to ELLs' academic language proficiency. ELLs need linguistic capital—the academic language needed to comprehend informational texts and speak and write what they know—in order to read complex informational texts; however, in reality, they do not have nearly enough. Fang (2004) also defined the importance of academic language as "power code," the language of power and privilege, to imply that individual ELLs cannot obtain academic and career success, without mastering the language of science (p. 343).

Research has identified several areas that negatively affect ELLs' informational text reading. First of all, science vocabulary, which is extremely abstract and technical in nature, poses the biggest problem for ELLs (Diaz-Rico and Weed 2002; Fang 2004). Words such as photosynthesis, stomata, oxidation, and tectonic are not only tongue twisters, but also conceptually complex for ELLs to use them in the correct context. Words with multiple meanings also make expository reading extremely difficult (i.e., power in *power* of speech versus disconnect the *power*, or plate in *plate*ful of food versus *plate* tectonics). Functional words (e.g., *therefore, while*, and *as*) that give information about rhetorical relationships such as cause/effect within the text can compound reading difficulties, and are therefore troublesome. Syntax—a structure of a sentence and grammatical rules of word order—also could make figuring out the meaning of a text even more complicated. The paragraph below, especially sentences two and three, demonstrates that the meaning of long compound-complex sentences are highly complicated and convey multiple layers of closely related information by expanded syntaxes. The compound-complex sentences typically consist of nominalized, subordinate, and/or coordinated clauses and phrases (Fang 2004). In other words, a subject and verb do not line up side by side, and simply figuring out the basic frame of the sentence becomes problematic. Along with content-dependent complexity, syntactic complexity adds an extra burden on ELLs when trying to comprehend written scientific discourse, to say the least.

> 1. In order for a solid to melt, the energy of the particles must increase enough to overcome the bonds that are holding the particles together (25 words). 2. It makes sense then that a solid which is held together by strong bonds will have a higher melting point than one where the bonds are weak, because more energy (heat) is needed to break the bonds (37 words). 3. In the examples we have looked at, metals, ionic solids and some atomic lattices (e.g. diamond) have high melting points, whereas the

melting points for molecular solids and other atomic lattices (e.g. graphite) are much lower (36 words). 4. Generally, the intermolecular forces between molecular solids are weaker than those between ionic and metallic solids (16 words). (Horner et al. 2008, p. 28)

In addition, the written discourse used for data analyses and argument construction in science requires metacognitive comprehension skills (Cavegnetto 2010). For instance, the task of data analysis asks students to establish relationships, explore cause and effect, and develop an argument to show that they can reason from evidence. It takes, however, ELLs' adequate reading comprehension strategies, such as chunking and outlining, to grasp argumentation discourse when dealing with text-driven expository nature of informational texts (Dobb 2004).

> The plant that received more light grew more (Claim). On average, for the six plants that received 24 hours of light, they grew 20 cm, had six yellow flowers, had fifteen leaves, and they were all bright green. On average, for the six plants that received 12 hours of light, they grew 8 cm, had two yellow flowers, and had four leaves. Also, two of the plants had zero flowers. These plants were still bright green, but they were smaller and with fewer flowers and leaves (Evidence). Plants require light to grow and develop. This is why the plant that received 24 hours of light grew more (Reasoning). (McNeill and Krajcik 2012, p. 37)

In science, charts, graphs, and pictures are frequently used to communicate major trends, relationships, and microscopic features of a scientific phenomenon. Visually presented information with such devices, however, can also be complex, and it requires a great deal of language competence to understand, interpret, and communicate what is presented (Lai et al. 2016). In addition to being cognitively demanding, comprehending informational text in science is linguistically demanding. The magnitude of the role reading plays in improving science learning must be acknowledged. The National Science Teachers Association's (NSTA) positional statement declares that science teachers should include opportunities for ELLs to develop science content literacy (NSTA 2009).

Method

This chapter reports a portion of the results from a state-funded grant program that aimed to increase high school science teachers' pedagogical knowledge in teaching CCSS-aligned informational texts through close reading to ELLs. It should be noted that this PD program did not attempt to increase teachers' particular science content knowledge. The objective of the PD was rather to provide the participating teachers with broadened perspectives about ELLs as learners and present a repertoire of new strategies regarding teaching science informational text reading to ELLs. This particular grant is a part of federally funded PD known as *Improving Teacher Quality (ITQ) Grant Program*, but is awarded by the state. The ITQ grant does not stipulate the program duration as long as the program consists of 45 contact hours; however, $ITS^4$4ELLS (**I**nformational **T**ext **S**pecialists in **S**ocial **S**tudies and **S**cience for **E**nglish **L**anguage **L**earners) was designed to last for an entire year to provide sustained support and PD environment for the teachers. $ITS^4$4ELLS had several program components: (1) three learning modules which introduced teachers to the unpacking of CCSS, second language acquisition theories and their applications in classroom discourse, and best practices on instructing science informational text reading to ELLs; (2) a teacher learning community (TLC), a learning unit that allows collaborative lesson studies for the participating teachers; (3) summer workshops; and (4) lesson studies conducted by individual TLC groups (i.e., after each module workshop, teachers collaboratively planned a lesson incorporating module content, and one teacher from each TLC volunteered to teach the lesson while the rest observed). Then the group got back together to review and discuss the teaching demonstration, modify and revise the lesson, and finalized it as the exemplary lesson for non-participating teachers. The following artifacts were collected: (1) interactive monthly reflections on the online forum, (2) teacher surveys at the completion of the program, (3) lesson plans, and (4) teacher interviews a year later. Five high school science teachers participated in the program (see Table 5.1).

Table 5.1 Participants

Participants	Gender	Years of teaching	Science level
Teacher 1	Female	8 years	High School
Teacher 2	Female	8 years	High School
Teacher 3	Male	3 years	High School
Teacher 4	Male	13 years	High School
Teacher 5	Male	19 years	High School

For the data analyses, all verbatim data were read multiple times by both authors, who agreed upon the key phrases generated from data-readings. There emerged six possible themes (more tools, update current strategies, my own understanding, better understanding of ELLs, good for everyone, and collaboration), which were narrowed down to five themes discussed in the Findings section below.

Findings

Teachers' Reconstructed Notions Regarding ELLs

Teachers' understanding of ELLs was very superficial at the beginning; however, upon completion of the program, it was reported that they gained significant knowledge about second language acquisition (SLA) and its impact on informational text reading. Research suggests that teachers might have erroneous ideas about ELLs (Brown 2007; Samson and Lesaux 2009; Harry and Klingner 2014). For instance, general education teachers may think that ELLs who are orally proficient would not have difficulties understanding science lectures. However, conversational English with friends in the hallway is qualitatively different from academic English required in science learning (Cummins 2000). The following verbatim text from participating teachers shows that they were deepening their understanding of ELLs as learners and what they need to do as teachers to help them succeed in content learning and informational text reading:

> We use a lot of phrases and words in our speech because they are common to us; we have heard them before. But to ELL students they won't have that background knowledge to really understand it. It really makes a difference in how I teach. I have noticed after this module [SLA], I am focusing more on vocabulary and looking for roots in words and trying to bring back an understanding from words we have previously learned in class. (Teacher 1)

She came to understand the complexity of the vocabulary issues faced by ELLs. That is, teaching the target concept at hand cannot be achieved if ELLs lack the background knowledge vocabulary that would facilitate them to comprehend the target concept. Difficulty of science learning is compounded by the interconnected nature of science as a discipline.

> The biggest thing that I got out of this training, breakthrough for me was that... I realized that I have to serve as a bridge between the terminology and the language that is not quite technical, not science language that is, but the higher level of English as opposed to the middle school or elementary level. I was great at teaching them science using the low level English. Tell them okay this is the scientific term, which puts the basic English at a higher level. Help facilitate some kind of bridge between the terminology that students understand and the language that the test banks use. The language of test banks is at higher levels and that shuts the students off. Me realizing that was a breakthrough for me. I started to think.... Hey, I need to take more steps to make build this bridge between the linguistic level of students and the language of tests. (Teacher 5)

He used the word, "breakthrough," to signify his realization in which in which ELLs' science learning is further exacerbated by the expository features of academic register or Tier 2 vocabulary, according to CCSS, characterized by formal language that is not domain specific. Before discussing differences between conversational English (e.g., Facebook English) and academic English (e.g., formal expression describing scientific phenomena), he had previously thought that unpacking of concepts with easier language would suffice; however, he realized that simplifying the concepts alone does not remove linguistic challenges for ELLs

> Second language acquisition module was wonderful... my Korean student is on Facebook. She can obviously speak English. Now I know she can speak in the lunch room but not in front of everybody about science in the science classroom. I didn't know it takes different times to learn different language. (Teacher 2)

She demonstrated greater understanding of differences between conversational English versus academic English by using her real-life examples for her ELLs. Even in andragogy, concretizing abstract notion is necessary if changes in the instructional behavior are the goal.

The verbatim reported here shows that these teachers recognized that the science discourse is more than just technical terminologies. After such realizations, it is more likely these teachers will scaffold the science register to reduce the language gap between their ELLs' current English and the academic English required in learning to explain science.

Refining Prior Knowledge and Strategies While Acquiring New Ones

Teachers indicated that the PD program helped them in two ways: they refined the strategies that they were already using and at the same time, they also acquired new ones of which they were convinced would make their instruction more effective for ELLs. All participating teachers indicated that they never knew that reading, along with vocabulary, could be chunked and broken down into meaningful parts to make reading passages more comprehensible to ELLs. Chunking a text is a powerful reading strategy in making challenging texts more manageable and reducing anxiety for readers (Brent and Millgate-Smith 2008). Such skills have to be explicitly taught to ELLs; thus, it is necessary for teachers to know about them (Chamot 2004). One particular teacher described how he would further build on the strategy he had used in the past:

> Never thought about having to break down reading and vocabulary... I'm learning new strategies that I have never tried before also realized that this can be applied to lower students in general. I feel that the close reading is something I can implement into my classroom now and would only need to slightly modify what we already do for annotation. It will help my students who struggle by re-reading the same article. (Teacher 3)

This is an example of which a participating teacher realized that some of the "extra steps" required to make informational texts more comprehensible do not even have to be elaborative. This is considered a meaningful finding, because it is critical for content teachers to understand that instructional strategies for ELLs could be as simple as breaking apart some vocabulary words or a text into several parts. If teachers perceive they already possess the pedagogical knowledge to accommodate ELLs, tweaking the existing knowledge is more "approachable," thus, "doable."

> Learning about second language acquisition stages was an eye-opener for me too. I also used annotating text during the lesson. Modeling rather than giving just instructions was more successful and less stressful/frustrating. (Teacher 4)

Stages of second language development are not difficult concepts to grasp. Once teachers realize how these language acquisition stages directly influence individual ELLs' learning in general, they can be a powerful tool; however, the challenge is that content teachers usually step into classroom without such knowledge leaving both teachers and students frustrated.

> I do a lot of visualization types of things. I think it's a skill that a lot of our low readers don't have. I found that students do not visualize what they read. One of my students said they saw nothing. They didn't have a movie in their mind. I thought everybody sees the movie like I do. So my lowest saw nothing. Low readers just see words only, intermediate readers saw in cartoon form, like Sponge Bob. Some saw a real snail in the woods or see the rivers. It's like a scale from none to cartoon to real. It is amazing. Now I do this all the time. I bring lots of pictures related to what they are about to read. I ask, what do you see? Tell me what do you see? Showing more

> visuals and tie with reading. Then, ask them do you remember a time when I showed you this picture? Try to connect what they are reading to the visuals. They seemed to enjoy more. There is a student who consciously trying to visualize. When they see more, they understand better. I found it enlightening that students may be challenged when they need to transition between text and figures and graphs when engaged with reading. Prior to this training, I have just taken this skill for granted and neglected explicitly teaching the skill only addressing it when modeling reading text with the class as a whole. (Teacher 2)

This teacher's account on how she discovered not every student saw "the movie" like she did is quite exciting because she had assumed that all students visualize what they read. Understanding what visualization does for reading comprehension, she deliberately increased her efforts to ensure her students visualize from reading and connect visual information back to the text. What is apparent is that she was fine-tuning and upgrading her current strategies to better accommodate ELLs. Such small changes can be led to tangible improvement in student learning if the teachers realize that they already possess the skills (Cazden 2001).

Construction of Teacher Knowledge

Most importantly, the participating teachers seem to personally have figured out how they would apply what they learned in the PD program into their daily practice. This is a very important aspect of a high-quality PD program. Without teachers "picking themselves up," nothing would really translate into the classroom-level practice. All the discourse of "science for all" would become rhetoric for ELLs, because when teachers are pressed for time, differentiating instruction does not become a priority (Walker et al. 2004). There was an abundance of evidence wherein teachers demonstrated personalizing and conceptualizing of close reading strategies for ELLs. Teachers' discourse in their monthly discussion indicated that they dug in to make sense of what they were learning, interpret it, and to create better instruction for ELLs. Teachers did not just regurgitate what was discussed during the workshops. They came up with ideas of their own for applying the newly learned strategies. For instance, one teacher suggested to others that they look at students' text books, not the teachers' edition, which is good evidence to show that teachers were critically thinking and applying the ideas of anticipating language difficulties for ELLs prior to teaching:

> I agree that by using the student text books we probably would be able to find more strategies to teach them more efficiently than by using the teacher's edition. We get to see what they see and are able to distinguish problems they face as they read the book. I hate the teacher's edition and rarely use the suggested materials/labs or questions they provide. (Teacher 4)

What is interesting about his comment is that the teachers were putting themselves in ELLs' shoes based on their newly acquired knowledge on SLA. During the workshops, many examples of science discourse were closely examined at the sentence level but the grade-level science textbooks were not used for that purpose. Discussion among the teachers about using the text book that students read, not the teacher's edition, illustrates that teachers were actively engaged in internalizing workshop content and extending knowledge base within their personal domain.

> I think it will be important to make sure that the materials we are offering for a close read are not beyond the frustration level of the students with whom we are working. This may involve some adaptation of materials or locating related content at a more appropriate level for close reading. The challenge (to implement CCSS in content class) is to find a way to add value and relevancy to the content we are covering. (Teacher 3)

When one of the workshops dealt with SLA theories, Krashen's $i+1$ (Krashen 2003) concept was introduced to highlight how challenging grade-level informational texts are and grossly beyond their current level of English. This teacher added the role of frustration in reading to the idea of comprehensibility.

> I felt like my first CC lesson went well. We spent a class period with a close reading and then a discussion of the vocab words that the kids didn't know. I was amazed at the vocab words the students didn't recognize. We

> then spent another class period with text based questions in which the students had to gather evidence from the text in order to support various statements. I then tried to get them to apply the knowledge learned from the lesson to other scenarios. I have to be more explicit with my instruction. Many didn't get the connection right away, so I had to reteach some of the material in more general terms. (Teacher 2)

After her teaching demonstration in front of her PD peers, she realized that she had to be even more explicit than she had originally planned, indicating that it takes some practice to link conceptualized knowledge and instructional delivery.

There came a very interesting yet profoundly noteworthy moment when one of the teachers posted the reflection below. His comment sets him apart from other teachers, because he acknowledged that in reality, extra scaffolding for ELLs takes time on the front end, but the lost time can ultimately be made up because he might not have to reteach. This acknowledgement is remarkable, for he appeared to be internalizing that well-conceptualized learning can be accelerated, but the care has to be taken at the beginning of the learning process. One of the reasons general education teachers give for their resistance to accommodating ELLs is a lack of time (Walker et al. 2004). Time is a zero-sum game and teachers are under pressure to cover the mandated curriculum. Yet, accommodating ELLs' learning needs means more frequent, in-depth scaffolding, which takes time and effort (Brown & Broemmel 2011). This teacher demonstrated deeper conceptualization of the dynamic concept of differentiated instruction for ELLs.

> Ultimately, I would hope that the extra effort to back up and meet them where they are at the initial sacrifice of addressing the standards in the most expedient manner—might build skills so that the time could be recovered later (after internalization) if the later standards are grasped faster and with more surety due to the new lens through which the students hopefully will become accustomed to using. (Teacher 3)

Albeit short, this reflection underscores the importance of teachers constructing their own knowledge if PD is to be translated into changed instructional behavior.

Utilitarian Perspectives: ELL-Strategies Are Good-For-Everybody-Strategies

Participating teachers understood that the strategies that work for the ELLs would help the rest of the general education students who are just struggling readers. This is a meaningful finding, because general education teachers tend to think that instructional accommodations they have to provide for ELLs are something extra they have to perform since they pertain to only ELLs (Reeves 2004, 2006). If the general education teachers realize that what benefits ELLs also benefits their peers, it is more than likely that they will try accommodation strategies for ELLs. This might be a "backdoor" means of convincing general education teachers; such buy-in from teachers is what ESL educators need. Without the teacher buy-in, the instructional behavior that correlates to improved student learning might not be possible (Turnbull 2002).

> I believe the integration and implementation of ELL strategies can be beneficial for ALL students in our classroom. Low level students can definitely see gain from the EL strategies as often they are students with very low English skills verbally, written, and oral. The skills taught to us have really helped me be able to instruct them more efficiently and help them improve their reading abilities. These skills, while designed for ELL students will help all students. (Teacher 1)

She used the word *believe*, which connotes enthusiasm, to describe the usefulness of ESL strategies and her commitment to use them although not explicitly stated.

> I have found that the close read works wonderfully with my low readers in my Biology AP class. At first there were gripes and complaints. "We have to read this again?" was often said the first couple of times, but now we have done it enough that they are used to it. I feel it also helps all students I teach because often by the end of the page they don't remember what they just read. While frustrating for me as an adult, I remember doing the same thing is school at times. More often than not we re-read the same article/paragraph and then we discuss it. I often pair this with a text walk

> and point out vocabulary words they may find difficult or not have heard before. Especially with my low readers, these techniques help the students a lot. (Teacher 3)

He reported that close reading strategies worked "wonderfully" with his AP students and also provided concrete examples of when exactly and how he used close reading strategies. If a teacher finds a way to implement ESL strategies learned in a PD program and has a success story to share, it is highly likely that he or she will be steadfast in accommodating ELLs.

> I also think that some of the strategies might even be of use to help prime my AP students for better integration of the volume of information that they must manage—often largely on their own—to allow for the lab time needed for greater focus on inquiry. (Teacher 3)

Participating teachers reported here reasoned that ESL strategies are good for all students, thus, useful. Convincing content teachers with an idea that accommodating ELLs is time well spent may be highly important since research shows that general education teachers are reluctant to address learning needs of ELLs due to time constraints (Reeves 2006). ESL educators rather want to present a moral argument as to why we, as a society, have to educate them in the first place; however, if this utilitarian perspective can convince content teachers to use ESL strategies, that is a step forward in making inclusive instructional practice universal.

Spread the Good: Reach Out to Fellow Teachers

The teachers were interviewed a year after the grant program was completed. It was rather gratifying to hear that the participating teachers reported that they still put into practice what they had learned in the PD program. The ultimate goal of the PD is to have teachers engaged in sustained practice in their classroom (Sparks 2002). The teachers reported that, in their science classes, they consistently implemented informational text reading strategies through close reading, and raved about the positive impact from such practices. One of the teachers shared a quite extraordinary story during the year-later interview. She mentioned that she started a conversation with teachers in the lunchroom since she found the usual lunchroom conversation unpleasant:

> I talked about what we learned and what has worked in my teaching. It keeps you from gossiping. My neighbor would say, "Can you go over close reading again with me?" Now I'm the designated teacher for ELLs. I'm the ESL expert (laugh). Math has been on board with close reading because I lunch with them (laugh). The administrators saw math struggling with informational texts and science does well with informational texts. I did presentations for all science teachers as well. The administrators asked me to do an in-service workshop for the entire school faculty... I think we should do this together. Why are we doing separately? And we started talking like let's do something together: Can we do a whole new unit, a huge unit... A math teacher suggested that we do close reading aligned with math and science lab. In math, they teach how to figure out the slope. In science the slope means something like acceleration... science is a real world. It is about how to apply to the real world. You are doing this it's happening in the real world (Teacher 2).

She not only taught other science teachers who did not participate in the program about informational text reading, but as shown, she even reached out to math teachers who lunched together in the teachers' room. Another teacher also shared the similar story.

> Our department is wonderful about sharing good ideas. I have been sharing with the others. I have told them about close reading. It usually starts with crying in the hallway. A bunch of crying because teachers are a bit frustrated because they are not getting through what they need to. We brainstormed as a department. I tell them how I use close reading strategies. Talk about what works in my class and what they might want to try. I talk about pre-teaching vocabulary, it's incredibly essential. I have watched teachers teach the vocabulary after they did reading. I understand what they were trying to do. They try to get the context clues of what the

> vocabulary words were before, but after seeing how much easier if we pre-teach vocabulary... then, you can incorporate context clues and talk about them. It is kind of hitting a brick wall. It doesn't work as well (Teacher 3)

Again, what has been conveyed from these teachers' interviews is that the impact of ITS44ELLS has been far reaching beyond the participating teachers. It goes to show that a high-quality PD could have sustaining power that encourages teachers to stay on the path to inclusive classroom practice.

Discussion

This chapter reports the results of a PD designed to improve high school science teachers' pedagogical knowledge related to close reading of informational texts by ELLs. As discussed in the findings, all participating teachers ubiquitously mentioned how they gained more the tangible tools in their strategy toolbox that they could use to promote ELLs' learning. This obvious statement sounds rather oversimplified; however, it implies that science teachers do not have all the necessary tools to be able to successfully teach informational text reading to ELLs. Teachers also reported that they not only gained new strategies, but they also retooled their existing ones to focus on ELLs to develop reading skills. Teachers were certainly enlightened after learning closely about how SLA affects teaching informational texts. They also understood the kinds of extra steps they have to take beyond their newly developed empathy.

The teachers also had their "aha moments" in realizing that their role goes beyond that of traditional science teachers in the era of Common Core; they must add on language scaffolding for better science learning. They deepened understanding about the complex nature of learning a subject like science in a weaker language, especially in informational text reading. Findings from this PD program also support the notion that teachers not only have to buy into the idea of putting in more effort for better results, but eventually they have to interpret, build, and reconstruct the professional knowledge coming their way (Darling-Hammond and McLaughlin 1995). Participating teachers in ITS^{4}4ELLS demonstrated that they put their "personal spin" to make sense of what they were learning and figure out how realistically they could implement ESL strategies in their own classroom.

One of the most significant findings from this project is participants reported that they sustained practice after a year. Research shows that teachers do revert back to their old ways of teaching (Guskey 2002; McIntyre and Kyle 2006). Contrary to such a fact, these teachers continued to pre-teach vocabulary, connect it to the text, visualize the reading, and annotate the informational text. More profound teacher change was evident in that they "converted" other teachers in their departments and even reached different departments. One of the hallmarks of an effective PD is teacher change and sustained practice (Whitworth and Chiu 2015), and ITS^{4}4ELLS confirmed that high-quality PD programs not only induce pedagogical shift in instructional practice, but also spread beyond the participating teachers' boundaries. These findings are closely in line with literature indicating that a high-quality PD program contributes to teachers' imminent need to develop more advanced and specific skill sets to meet ELLs' academic learning needs. While top-down educational initiatives do not actually provide classroom teachers with opportunities to develop professionally, these results draw attention to the need for ongoing support for teachers who must prepare ELLs to be college- and career-ready.

Preparing teachers to meet the needs of ELLs is a must-goal, not a recommended goal or a good-to-have-goal. There is no alternative to this goal; however, it has been consistently difficult to achieve. Based on this small-scale PD study, we can be optimistic about teacher change. When we reach out to teachers and provide them with the knowledge, skills, and strategies they need but do not receive elsewhere, teachers are much more willing to transform themselves. The key, however, is for districts and states to invest in teachers' continuing PD. ITS^{4}4ELLS allocated $2,000 for substitute teachers and a $900.00 stipend for each teacher, almost a $3,000 budget per teacher. Perhaps the positive results are related to the indirect effect of the teachers' feeling appreciated as professionals by the PD program. High-quality PD programs require financial resources, but it might be the only way to achieve the goal that we cannot afford to avoid. Educators still have much work to do to make inclusive practice a reality for ELLs, because there are many more teachers to be enlightened. Reeves' study (2004) below illustrates the discrepancy of the general education teachers' reported interests in learning about working with ELLs and actual commitment to their PD:

> In my survey of Eaglepoint's (research site) faculty, 51% of teachers agreed with the statement, "I am interested in receiving more training in working with ESL students," and 93% reported they had received no such training, but the only Eaglepoint teacher to attended the in-service (ESL-related) was Linda, the ESL teacher. Teachers' lack of attendance and their lukewarm interest in training can likely be attributed to a number of factors including the troubled history of one-shot in-service programs (Guskey and Hubarman 1995). Clair's (1995) research, however, suggest that general education teachers may feel that no special training is necessary for teachers to work successfully with ELLs. Eaglepoint's school-wide endorsement of equal treatment would support this assertion. (p. 51)

In closing, discussion of the limitation of the program is in order. First of all, all instructional behavior or perceptional changes discussed are based on teachers' self-reports from interviews, and monthly online reflections. We must acknowledge that the teachers accounted positive experiences regarding ITS^{4}4ELLS while being reticent about challenges and concerns they had. They did not specify any personal struggles they went through while implementing informational texts reading strategies; perhaps they were being gracious to funders. If they had to overcome any challenges during the conceptualization process internally or externally, those are unknown to us. Accordingly, findings from this PD program do not add different dimensions to the existing literature in that regard. Another shortcoming of this PD program is that the direct impact of this particular program was not quantified, as findings reported in this chapter are not triangulated by students' standardized test scores limiting the interpretations of the program findings. Research does show teachers' reported changes in their practice following PD, but does not necessarily reflect profound changes in their instructional behavior

(Weiss and Pasley 2006). In order to connect the validity of this PD program to increased student learning, we would like to follow up with these teachers as a next step to see whether the positive results correlate to ELLs' higher achievement on science tests. Albeit lacking such evidence, this PD endeavor goes to show that teacher success has to be a precursor to student success in that teacher success greatly matters.

References

Ball, D., & Cohen, D. (1999). Developing practice, developing practitioners: Toward a practice- based theory of professional education. In L. Darling-Hammond & G. Sykes (Eds.), *Teaching as the learning profession: Handbook of policy and practice* (pp. 3–32). San Francisco: Jossey-Bass.

Bourdieu, P. (1977). *Outline of a theory of practice.* Cambridge: Cambridge University Press.

Bourdieu, P. (1986). The forms of capital. In J. G. Richardson Ed., *Handbook of theory and research for the sociology of capital* (pp. 241–258). New York: Greenwood Press.

Brent, M., & Millgate-Smith, C. (2008). Reducing anxiety and increasing motivation. In V. Camberwell (Ed.), *Working together for adolescents with a language learning disability* (pp. 7–20):). Australia: ACER Press.

Brown, B. A., & Ryoo, K. (2008). Teaching science as a language: A "content-first" approach to science teaching. *Journal of Research in Science Teaching, 45*(5), 529–553.

Brown, C. L. (2003). Who is responsible for English-language learners? A case study from a third-grade classroom. *Academic Exchange Extra(March).*

Brown, C. L. (2007). Content-based ESL instruction and curriculum. *Academic Exchange Quarterly, 11*(1), 114–119.

Brown, C. L., & Broemmel, A. (2011). Deep Scaffolding: Enhancing the reading experiences of English language learners. *New England Reading Association Journal, 46*(2), 34–39.

Cavagnetto, A. R. (2010). Argument to foster scientific literacy: A review of argument interventions in K-12 contexts. *Review of Educational Research, 80*, 336–371.

Cazden, C. B. (2001). *Classroom discourse: The language of teaching and learning.* Portsmouth, NH: Heinemann.

Chamot, A. U. (2004). Issues in language learning strategy research and teaching. *Electronic Journal of Foreign Language Teaching, 1*(1), 14–26.

Clair, N. (1995). Mainstream teachers and ESL students. *TESOL Quarterly, 29*(1), 189–196.

Coleman, J. S., Campbell, E. Q., Hobson, C. J., McPartland, J., Mood, A. M., Weinfeld, F. D., & York, R. L. (1966). *Equality of educational opportunity*. 2 volumes. Washington, D.C.: Office of Education, U. S. Department of Health, Education, and Welfare, U. S. Government Printing Office. OE-38001; Superintendent of Documents Catalog No. FS 5.238:-38001.

Cummins, J. (2000). Academic language learning, transformative pedagogy and information technology: Towards a critical balance. *TESOL Quarterly, 34*(3), 537–548.

Darling-Hammond, L. (1999). *Teacher quality and student achievement: A review of state policy evidence*. Seattle: Center for the Study of Teaching and Policy, University of Washington.

Darling-Hammond, L. (2000). Teacher quality and student achievement: A review of state policy evidence. *Education Policy Analysis Archives, 8*(1), Retrieved from http://epaa.asu.edu/epaa/v8nl.

Darling-Hammond, L. (2004). Inequality and the right to learn: Access to qualified teachers in California's public schools. *Teachers College Record, 106*(10), 1936–1966.

Darling-Hammond, L., Wei, R. C., & Andree, A. (2010). How high-achieving countries develop great teachers. *Stanford Center for Opportunity Policy in Education*. Retrieved from https://edpolicy.stanford.edu/sites/default/files/publications/how-high-achieving-countries-develop-great-teachers.pdf

Darling-Hammond, L., & McLaughlin, M. W. (1995). Policies that support professional development in an era of reform. *Phi Delta Kappan, 76*(8), 597–604.

Diaz-Rico, L. T., & Weed, K. Z., *The crosscultural, language, and academic development handbook: A complete K-12 reference guide*, 2nd. (Boston: Ally & Bacon, ed. 2002).

Dobb, F., *Essential elements of science instruction for English learners*, 2nd. (Los Angeles: California Science Project, ed. 2004).

Dorph, R., Shields, P., Tiffany-Morales, J., Hartry, A., & McCaffrey, T. (2011). *High hopes– few opportunities: The status of elementary science education in California*. Sacramento, CA: The Center for the Future of Teaching and Learning at WestEd.

Duke, N. K. (2004). The case for informational text. *Educational Leadership, 61*(6), 40–44.

Education Week Research Center. (2013). *From adoption to practice: Teacher perspectives on the Common Core*. Retrieved from http://www.edweek.org/media/ewrc_teacherscommoncore_2014.pdf

Fang, Z. (2004). Scientific literacy: A systemic functional linguistic perspective. *Science Education*, 89, *335–347*.

Fang, Z. (2006). The language demands of science reading in middle school. *International Journal of Science Education*, *28*(5), 491–520.

Fang, Z., & Wei, Y. (2010). Improving middle school students' science literacy through reading infusion. *The Journal of Educational Research*, *103*(4), 262–273.

Gallimore, R., Ermeling, B. A., Saunders, W. M., & Goldenberg, C. (2009). Moving the learning of teaching closer to practice: Teacher education implications of school-based inquiry teams. *Elementary School Journal*, *109*(3), 537–553.

Gee, J. P. (2005). Learning by design: Good video games as learning machine. *E-Learning*, *1*, 5–16.

Guskey, T. (2002). Professional development and teacher change. *Teachers and Teaching*, *8*(3), 381–391.

Guskey, T., & Huberman, M. (1995). *Professional development in education: New paradigms and practices*. New York: Teachers College Press.

Hanushek, E. A. (1992). The trade-off between child quantity and quality. *The Journal of Political Economy*, *100*(1), 84–117.

Hanushek, E. A., Kain, J. F., & Rivkin, S. G. (1998). *Does special education raise academic achievement for students with disabilities?* (No. w6690). Cambridge, MA: National Bureau of Economic Research.

Harry, B., & Klingner, J. K., *Why are so many minority students in special education? Understanding race and disability in special education*, 2nd. (New York: Teachers College Press, ed. 2014).

Hill, H. (2009). Fixing teacher professional development. *Phi Delta Kappan*, *90*(*7*), 470–477.

Horner, M., Halliday, S., Blyth, S., Adams, R., & Wheaton, S. (2008). *The free high school science tests: Textbooks for high school students studying sciences chemistry grades 10–12* J. Padayachee, J., Boulle, J., Mulcahy, D., Nell, A., Toerien, R., & Whitfield, D. (Eds.). http://nongnu.askapache.com/fhsst/Chemistry_Grade_10–12.pdf

Krashen, S. (2003). *Explorations in language acquisition and use: The Taipei lectures*. Portsmouth, NH: Heinemann.

Lai, K., Cabrera, J., Vitale, J. M., Madhok, J., Tinker, R., & Linn, M. C. (2016). Measuring graph comprehension, critique, and construction in science. *Journal of Science Education and Technology, 25*(4), 665–681.

Lee, O., Hart, J., Cuevas, P., & Enders, C. (2004). Professional development in inquiry-based science for elementary teachers of diverse student groups. *Journal of Research in Science Teaching, 41*(10), 1021–1043.

Lee, O. (2005). Science education and English language learners: Synthesis and research agenda. *Review of Educational Research, 75*(4), 491–530.

Lee, O., Quinn, H., & Valdés, G. (2013). Science and language for English language learners in relation to next generation science standards and with implications for common core state standards for English language arts and mathematics. *Educational Researcher, 42*(4), 223–233.

Lee, O., Miller, E., & Januszyk, R. (2014). Next generation science standards: All standards, all students. *Journal of Science Teacher Education, 25*(2), 223–233.

Lee, O., & Fradd, S. H. (1998). Science for all, including students from non-English language backgrounds. *Educational Researcher, 27*, 12–21.

Lemke, J. L. (1990). *Talking science: Language, learning and values.* Norwood, NJ: Ablex.

McIntyre, E., & Kyle, D. W. (2006). The success and failure of one mandated reform for young children. *Teaching and Teacher Education, 22*(8), 1130–1144.

McLaughlin, M. W., & Talbert, J. E. (2001). *Professional communities and the work of high school teaching.* Chicago: University of Chicago Press.

McLaughlin, M. W., & Talbert, J. E. (2006). *Building school-based teacher learning communities: Professional strategies to improve student achievement.* Chicago: University of Chicago Press.

McNeill, K. L., & Krajcik, J. (2012). *Supporting grade 5–8 students in constructing explanations in science: The claim, evidence and reasoning framework for talk and writing.* New York: Pearson Allyn & Bacon.

Miles, K. H., Odden, A., Fermanich, M., & Archibald, S. (2004). Inside the black box of school district spending on professional development: Lessons from comparing five urban districts. *Journal of Education Finance, 30*(1), 1–26.

Mizell, H. (2010). *Why professional development matters.* Oxford, OH: Learning Forward.

National Center for Education Statistics (NCES). (2002). *Schools and staffing survey, 1999–2000. Overview of the data for public, private, public charter, and Bureau of Indian Affairs elementary and secondary schools.* Washington, DC: U.S. Department of Education, Office of Educational Research and Improvement.

National Center for Education Statistics (NCES). (2012). *The nation's report card: Science 2011*. Washington, DC: U.S. Department of Education. Retrieved from http://1.usa.gov/1L5ug7b.

National Governors Association Center for Best Practices & Council of Chief State Officers. (2010). *Common core state standards*. Washington D.C.: National Governors Association Center for Best Practices, Council of Chief State Officers.

National Governors Association, & Council of Chief State School Officers. (2010). *Common core state standards*. Washington, D.C.: National Governors Association Center for Best Practices, Council of Chief State School Officers.

National Research Council. (2001). *Inquiry and the national science education standards*. Washington, DC: National Academy Press.

National Research Council. (2012). *A framework for K-12 science education: Practices, crosscutting concepts, and core ideas*. Washington, DC: The National Academies Press.

Reeves, J. (2004). "Like everybody else": Equalizing educational opportunity for English language learners. *TESOL Quarterly*, *38*(1), 43–66.

Reeves, J. (2006). Secondary teacher attitudes toward including English-Language Learners in mainstream classrooms. *Journal of Educational Research*, *99*(3), 131–142.

Rosebery, A. S., Warren, B., & Conant, F. R. (1992). Appropriating scientific discourse: Findings from language minority classrooms. *The Journal of the Learning Sciences*, *21*, 61–94.

Samson, J. F., & Lesaux, N. K. (2009). Language-minority learners in special education: Rates and predictors of identification for services. *Journal of Learning Disabilities*, *42*, 148–162.

Sparks, D. (2002). *Designing powerful professional development for teachers and principals*. Oxford, OH: National Staff Development Council.

The National Science Teachers' Association (NSTA). (2009). *Science for English language learners*. Retrieved (May, 2016) from http://www.nsta.org/about/positions/ell.aspx

Turkan, S., de Oliveira, L. C., Lee, O., & Phelps, G. (2014). Proposing a knowledge base for teaching academic content to English Language Learners: Disciplinary linguistic knowledge. *Teachers College Record*, *116* (3), http://www.tcrecord.org/library ID Number: 17361.

Turnbull, B. (2002). Teacher participation and buy-in implications for school reform initiatives. *Learning Environments Research*, *5*, 235–252.

Walker, A., Shafer, J., & Liams, M. (2004). "Not in my classroom": Teachers' attitudes towards English language learners in the mainstream classroom. *NABE Journal of Research and Practice, 2*(1), 130–160.

Walqui, A., & Van Lier, L. (2010). *Scaffolding the academic success of adolescent English language learners: A pedagogy of promise.* San Francisco, CA: WestEd.

Weiss, Iris R. and Pasley, Joan D. (2006). Scaling Up Instructional Improvement Through Teacher Professional Development: Insights From the Local Systemic Change Initiative. CPRE Policy Briefs. Retrieved from http://repository.upenn.edu/cpre_policybriefs/32

Whitworth, B., & Chiu, J. (2015). Professional development and teacher change: The missing leadership link. *Journal of Science Teacher Education, 26*, 121–137.

Yoon, K. S., Dunca, T., Lee, S. W.-Y., Scarloss, B., & Shalpley, K. (2007). *Reviewing the evidence on how teacher professional development affects student achievement (Issues & Answers Report, REL 2007-No.033).* Washington, DC: U.S. Department of Education, Institute of Education Sciences, National Center of Education Evaluation and Regional Assistance, Regional Educational Laboratory, Southwest.

Zwiers, J. (2007). Teacher practices and perspective for developing academic language. *International Journal of Applied Linguistics, 17*, 93–116.

Clara Lee Brown is Associate Professor of ESL Education in the Department of Theory and Practice in Teacher Education at The University of Tennessee, Knoxville. She is Program Advisor and Coordinator of the ESL Education program. Her research interests include enhancing ELLs' academic language in content areas, equity issues in assessments, and bilingual identity. She has widely published on issues regarding teaching ELLs in content areas and developing their academic language. She has implemented professional development grant projects for middle school math teachers and high school social studies and science teachers focusing on helping ELLs with content literacy.

Mehmet Aydeniz is Associate Professor of science education at The University of Tennessee, Knoxville (UTK), USA. He also directs the STEM GIFTED education program at UTK. Dr. Aydeniz's research focuses on pre-service and in-service science teachers' adoption of reform-based instructional practices such as argumentation and assessment of different interventions on student learning and their appropriation of epistemic practices of science. He taught high school chemistry in Seminole County, FL before joining UTK as a faculty in 2007.

6

Supporting English Language Learners in Secondary Science Through Culturally Responsive Teaching

Gretchen Oliver

Schools have a responsibility to provide a high quality education for all students, helping them to meet rigorous academic standards, specifically the Common Core Learning Standards in English Language Arts and Mathematics, as well as the Next Generation Science Standards. Yet, for culturally and linguistically diverse students, the demands of learning both grade-level content and academic English are compounded compared to their typical native English-speaking peers. According to Wepner (2011), an overhaul of the current educational system and a change in perspective are needed to respond to the new needs of schools.

If the goal in educating children, both English language learners (ELLs) and those who are native English-speaking (NES), is to prepare them to become full participants in U.S. economic, political, and cultural life, then providing the former with teachers who are well-versed in Teaching English to Speakers of Other Languages (TESOL) methods

G. Oliver (✉)
Department of Educational Theory and Practice, State University, Albany, USA
e-mail: goliver@albany.edu

© The Author(s) 2017

L.C. de Oliveira, K. Campbell Wilcox (eds.), *Teaching Science to English Language Learners*, DOI 10.1007/978-3-319-53594-4_6

represents the best hope for achieving this goal (Tèllez and Waxman 2006). Furthermore, if the teacher is the key to student learning, as much as of the research shows (Wardle 1996; de Jong and Harper, 2005; Knight and Wiseman 2006; Tèllez and Waxman, 2006; Walqui and Van Lier 2010; Oliver and Oliver 2013; Ramirez and Jimenez-Silva 2014), then teacher preparation and professional development (PD) programs must address the intercultural skills, knowledge, and instructional practices which all teachers, language specialists, and content-area teachers alike must possess to meet the needs of culturally and linguistically diverse students. It is not enough for mainstream teachers to simply employ "just good teaching" practices (de Jong and Harper 2005) in their instruction. Rather, teachers of ELLs must first work to meet their unique needs, and thereafter, all students will benefit. To meet these needs, Wardle (1996) advocates an anti-bias and ecological model for a multicultural education. In this model, the child is the focus, existing in his/her dynamic context, bringing with him/her a variety of experiences from which to draw in his/her learning. Moreover, Wardle (1996) explains that "educators should not assume what they teach is more important than how they teach" (p. 153), placing an equal emphasis on both content and pedagogical knowledge.

Such content and pedagogical knowledge is not distributed equally for all students. In fact, the achievement gap between ELLs and NES students in science demonstrates that ELLs do not have the same opportunities to engage with the content as non-ELLs, as expressed in the Next Generation Science Standards (NGSS Lead States 2013). According to the 2009 National Assessment of Educational Progress (NAEP), the gaps in science achievement among ELLs and their NES peers has widened, which is a reversal of the trend from the previous decade (National Center for Education Statistics 2009). While science educators can offer students concrete experiences through conducting experiments and observing natural phenomena, science is a language-based discipline, and one that employs a wide range of general- and content-specific academic language, technical terms, oral and written instructions for conducting experiments, and procedures for report writing (Short et al. 2013). Thus, the linguistic demands of this content area present challenges for ELLs.

One of the ways in which educators are beginning to address the achievement gap in science is through culturally responsive teaching, an approach which values the cultural and linguistic diversity that ELLs bring to school with them, and one that allows students to use all of their linguistic and cultural resources in the classroom. Such educators value the diversity ELLs bring to the classroom, and utilize it to help students maximize their learning potential. This chapter will recount the ways in which one suburban school district in New York State has provided ESOL-focused (English as a Second or Other Language) PD to its teachers, and how two secondary science teachers have implemented what they have learned into their practice. Moreover, it will show how this ESOL-focused PD has translated into culturally responsive teaching practices, influencing ELLs' learning, both in terms of developing their academic English and knowledge of science.

Theoretical Foundations

The existing literature provides many examples of how teachers and researchers alike have approached the education of culturally and linguistically diverse students. In addition, the education of these students can be viewed through a socio-ecological lens, one that builds upon the theory that that classroom instruction has a direct relation to its building and district support systems (Wilcox et al. 2015). Both will be explored in this section.

To begin, educating culturally and linguistically diverse students requires a multicultural approach to teaching, acknowledging the various resources and "funds of knowledge," the resources and lived experiences (González et al. 2005), that English language learners bring to the classroom with them. Simply put, ELLs are not empty vessels, coming to our school system without prior knowledge or experiences. Rather, they have unique strengths that teachers can use to tap into in their instruction. One feature of a multicultural approach is culturally responsive teaching, which rejects a deficit view of these students, moving away from a one-way accommodation, where students are viewed as lacking, inferior, and/or deprived (Nieto 2000). Instead, culturally responsive teaching provides a mutual accommodation situation, where teachers and other school professionals are able "to recognize and build upon the resources and assets that [culturally and linguistically diverse] students bring to the school. These resources, when used as a basis for instruction, enhance students' capacities to reach academic success" (Committee on Multicultural Education 2002).

In terms of PD for teachers of ELLs, Jiménez et al. (2015) outline a three-pronged framework for culturally responsive teaching. Herein, the focus is on teachers' dispositions towards ELLs, their knowledge of content and how to deliver it to students in ways that they can understand it, tapping into their funds of knowledge, and the use of students' native language (L1) and its relationship to the new language (L2), and viewing students as emergent bilinguals or even multilinguals. This type of PD supports the notion that teachers should consider the strengths and unique skill set ELLs bring to the classroom and to find ways to leverage these strengths and skills in ways that maximize their potential to be successful in the classroom.

Before ELLs enter mainstream classes, they may be provided with instruction in a sheltered environment to prepare them for the content and language they will encounter in classes with NES students. Sheltered instruction refers to separate classes that do not include NES students, as opposed to mainstream classes that include both ELLs and NES students (Freeman and Freeman 1988). The literature provides many examples of the benefits of both sheltered (e.g., Verma et al. 2008; McIntyre et al. 2010; Peercy 2011) and mainstream inclusion classes (e.g., Hansen-Thomas 2008; Theoharis and O'Toole 2011), and in many instances, one approach is favored over the other. Regardless of the instructional model (sheltered or mainstream inclusion), the existing scholarship also advocates PD for teachers of ELLs, assuring that they can make the content accessible for them, and to do so in a way that respects and values their prior experiences (Oliver and Oliver 2013; Jiménez et al. 2015). Additionally, the literature demonstrates that teachers need to be able explicitly to teach academic vocabulary and to allow students to use their native language as a way of supporting their learning (Gebhard and Willett 2008; Peercy 2011; Ramirez and Jimenez-Silva 2014). Finally, the literature demonstrates that taking a culturally responsive approach to teaching is a way to assist English language learners in becoming independent learners (Ramirez and Jimenez-Silva 2014) as it provides ELLs with access to grade-level content through rigorous instruction, and with the same expectations as their NES peers. It values students' resources, prior experiences, and native language and culture. Further, this approach promotes a sense of belonging to the greater school community.

This chapter is theoretically informed by Bronfenbrenner's (1977) theory of human development. Central to this theoretical perspective is the notion that the environment is made up of structures, and each individual structure is embedded in the next. This particular lens is fitting, because it can be used to explain how classroom instruction is connected to building and district systems of support that surround it (Wilcox et al. 2015). This theory posits the idea that classrooms are a part of a larger system, one that contains multiple layers, each influencing one another in a reciprocal manner. A classroom is nested within the larger systems of the environment, namely a building and a district. Moreover, these two systems are also nested within and influenced by the outside world.

Bronfenbrenner (1977) describes each structure as a microsystem, a mesosystem, an exosystem, and a macrosystem. In the context of this study, today's learning and teaching environment can be viewed in these terms: beginning with the macrosystem, there is the outside world which has an indirect impact on teaching and learning that takes place in a given schools. This may include national and state policy, historical, cultural and political systems, the local community, family and home life, and any outside partners in education. Moving inward, to the mesosystem, the district may (in theory) have an explicit vision and a mission vis-à-vis teaching and learning for students, and as such, its leaders develop and implement a variety of initiatives and program models to support this mission and vision.

Next, at, the exosystem layer, there is the building leadership, embodied in someone who might also have a personal vision and mission for teaching and learning, and who carries out the district's vision and mission. Additionally, these leaders implement program models, policies, and practices which impact teaching and learning. It should be noted that a leader's personal vision and mission for teaching and learning is not always in alignment with the scripted vision and mission of a school or district, and at times, the two might be at odds. Finally, at the microsystem level, within each classroom there is a specific dynamic between teaching and learning, between teachers and students, which is the result of the influences from the aforementioned systems.

Wilcox et al. (2015) differentiate between proximal influences (direct interactions between teachers and students at the classroom level) and those that are distal (outside of the macrosystem, yet have an impact on this system). Moreover, they argue that schools are complex systems; they have diverse student populations, teachers who come from a variety of teaching and learning backgrounds, and leaders who have personal visions for teaching and learning. Additionally, Portin (2004) notes that school leaders are affected by the actions of superintendents, school boards, and central offices. These individuals must comply with national, state, and local policies, as well as negotiate their way through historical, political, and cultural influences. Finally, Murray (2009) asserts that "while global issues impact language and teaching programs, all leadership is local in that it needs to be responsive to and support and sustain the environment in which the leadership occurs" (p. 14). In other words, leaders have to make decisions with an eye to the current national and/or state policy, and they must do so keeping in mind their context, including its needs, challenges, and assets. Moreover, both proximal and distal forces impact teaching and learning. The following graphic can be used to conceptualize an ecological model for teaching and learning, as well as the relationship between the various systems within this model (Fig. 6.1):

This graphic shows how the three sub-systems are nested within the larger macrosystem, where the realities of the outside world influence perspectives, practices, and products of the meso-, exo-, and microsystems. There is an interrelationship between the three sub-systems; the broken lines demonstrate how the different aspects of the macrosystem and the various sub-systems, although distal in nature, can permeate the borders of a classroom, building, and/or district, thus impacting teaching and learning. Finally, it should be noted that learning is a social process, and the environment in which a person lives has a direct influence on his/her actions, development, and perceptions (Piaget 1936;

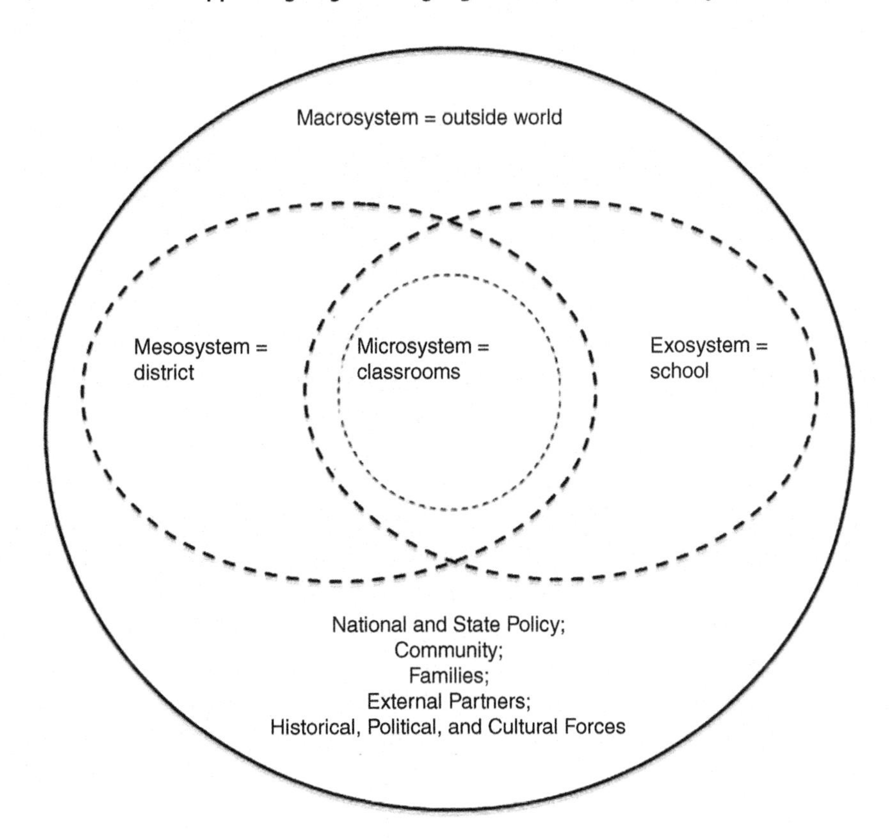

Fig. 6.1 Ecological model

Dewey 1938; Freire 1970; Vygotsky 1978, 1986; Lave and Wenger 1991). As such, this framework has been employed in an effort to describe the influence of culturally responsive teaching on English language learners in the secondary science classroom.

Method

This study employs a descriptive, single-case study approach (Yin 2014) to investigate the experiences of leaders, teachers, and students in one school in a real-life context. The rationale for employing a single case study is that this is a common case, "capturing the circumstances and conditions of an everyday situation" (Yin 2014, p. 52). The unit of analysis is a suburban school district and its high school, with focal subunits that include two teachers, one dually certified ESOL and Biology teacher, and another who is certified in Biology and Chemistry. In this school, culturally responsive teaching is ostensibly in place to support its English language learners as they work to meet both content and language proficiency standards, with the ultimate goal of achieving academic success as evidenced through graduating from high school, and even pursuing a post-secondary degree for others. The district provides structured and targeted PD to promote and reinforce this approach.

The focal subunits were purposefully selected because this district employs a content-based instruction approach in its ESOL program, and the study of science for ELLs is scaffolded in a deliberate sequence, in an effort to meet students where they currently are with their language proficiency, prior schooling, and experiences with science. This study employs an embedded case study design, looking both at the larger unit (the school/district) and at each subunit (an ESOL/Science teacher and a mainstream Living Environment teacher). The embedded design provides "significant opportunities for extensive analysis, enhancing the insights into the single case" (Yin 2014, p. 56). Equal attention will be given to the larger case and the subunits.

Context

Central Bridge High School (pseudonym) had approximately 1500 students enrolled in the 2014–2015 academic year, and 8.3% of the student population was identified as English language learners. The number of ELLs in the high school increased by 50% in one academic year, which was characterized as a "dramatic increase" by district personnel. Many of these students are SLIFE, Students with Limited, Interrupted, or Inconsistent Formal Education, who, according to the website ¡Colorín Colorado!,[1] have to learn not only academic content and language but also traditional classroom behaviors.

[1] ¡Colorín Colorado! is a bilingual resource site for families and educators of English language learners (www.colorincolorado.org).

The ESOL program at Central Bridge serves a range of students, with the majority coming from Central America (mostly Guatemala) and South America. At the high school, beginning and intermediate level students receive sheltered instruction in language arts, literacy, social studies, and science. Dually certified ESOL teachers teach sheltered courses in social studies and science for content-area credit. ESOL teachers and mainstream teachers collaborate and co-teach English Language Arts classes.

At the high school, they have developed the course sequence outlined below to help students to access science content while developing language skills (Fig. 6.2):

At the high school, beginners and newcomers start with ESL Science Literacy. They move on to ESL Science in 10th grade, a class where intermediate-level students continue developing their language abilities while learning living environment content. Intermediate-level 9th grade students may also take this course if they have successfully completed the ESL Science sequence at Central Bridge Middle School. Thereafter, students take a mainstream Living Environment Regents class, earning the Science credit they will need for graduation. In the past, students had been clustered and placed with a science teacher who works closely with the school's instructional assistant or with a science teacher that has

ESL Science Literacy
ESOL teacher
Beginning (ESL 1)
Grade 9/newcomers

ESL Science
Dually certified LE/ESOL teacher
Intermediate (ESL 2)
Grades 9-10

Living Environment Regents
Science teacher with instructional assistant
Grades 10-12

Fig. 6.2 Science-ESOL course sequence at central bridge

extensive experience working with ELLs. In 2014–2015, due to the implementation of a new building bell schedule, ELLs taking living environment classes were scattered across the many sections offered, thus providing challenges to a cohort model for these students.

Participants

Participants include representatives from the meso-, exo-, and micro-systems of the ecological model. Each was interviewed privately and was given a pseudonym. Table 6.1 provides information about their current and previous positions in the district.

Additionally, eight ELL students from the Living Environment and ESOL Science classes, ranging from beginning to advanced proficiency levels, have been included in this study. They were observed in classes, as well as interviewed in three groups, according to proficiency level (beginning, intermediate, advanced). Some of the students have taken courses in

Table 6.1 Participants

Participant	Current position	Years in current position	Previous position(s) in district
David	Assistant Superintendent for Curriculum & Instruction	8	Science Teacher; Coordinator of Science & Technology, 6–12
Vivian	ESOL Director, K-12	5	ESOL Teacher
Ben	High School Principal	6	N/A (came to district six years ago from another high school principal position)
Paul	Coordinator of Science & Technology, 6–12	3	Science Teacher
Anita	Sheltered ESOL Science Teacher	6	Co-teacher with Lisa for Sheltered ESOL Science classes
Lisa	Mainstream Living Environment Teacher	11	Co-teacher with Anita for Sheltered ESOL Science classes

the entire course sequence, beginning at Central Bridge Middle School, while others are newcomers and in their first class at the high school.

Data Collection and Analysis

With the goal of developing an in-depth understanding of this case (Creswell 2013), multiple forms of data were collected over a 16-month time frame through observations and direct contact with participants. These data include documentary evidence (including teacher-written needs assessment and end of the year summative reports, lesson plans, teacher reflections, district publications, demographic data, and information on website), direct observations of classes (using field notes and videos), and meetings and interviews with individuals from each of the "systems." A case study database was created to file and archive all evidence, and a case study protocol was employed in an effort to stay consistent with collection and filing of data, and thus creating a chain of evidence.

Using an embedded case study design (Yin 2014), analysis focused on both the district and school as a unit, as well as the ESOL and Science classes as subunits. The goal in the analysis was to develop a case description, as it will help "to identify the appropriate explanation to be analyzed" (Yin 2014, p. 140). *A priori* codes from the literature review were identified and the data were systematically categorized through qualitative data software, NVivo, according to the theoretical framework, using each of the four systems from the ecological model.

Following Miles, Huberman, and Saldana's (2013) "Methods of Describing," these data were displayed though conceptually clustered matrices. Through rows and columns, these matrices provided "an at-a-glance summative documentation and analysis" (p. 173). This procedure provided a means of triangulating the data, seeing patterns and relationships, as well as noting convergences and divergences among the various informants and data sources. Finally, in order to build added validity and trustworthiness, drafts of this study were shared with participants in an effort to verify its trustworthiness though member checks.

Results

Data analysis provided evidence of the presence of many aspects of culturally responsive teaching at Central Bridge. The most prevalent and those which corresponded to the existing literature include: PD that promotes culturally responsive teaching, differentiation, teacher dispositions, the student as the focus, and ELL students' access to programs. For the purposes of this chapter, PD and differentiation will be discussed with explicit examples from the various data sources.

Professional Development and Culturally Responsive Teaching

At Central Bridge, there is a strong focus on PD for all teachers, much of which has an ESOL focus, promoting a culturally responsive instructional environment. In each of the six interviews with administrators and teachers, participants referenced access to meaningful PD as being integral to both teaching and student learning. First, at the district (meso-) level, as part of the three year induction process, all teachers participate in job-embedded workshops related to Saphier's *The Skillful Teacher*,[2] the SIOP®[3] Model of Instruction, and LATIC[4] (learner-active, technology-infused classroom) training. David views these programs as a means for any teacher to develop a "deeper repertoire for working with English language learners." Moreover, both David and Ben have participated in the same SIOP®

[2] The Skillful Teacher and its parent organization, Research for Better Teaching, provide a professional development framework based on two core principles: "every child deserves a quality education regardless of the circumstances of their birth; and all children are capable of growing their ability and learning" (http://www.rbteach.com/). Central Bridge is one of many districts in the country who have adopted this professional development model as a systematic approach to school improvement.

[3] The Sheltered Instruction Observation Protocol (SIOP®) Model is a lesson planning and delivery model of instruction as well as an observation instrument for rating the fidelity of lessons to the model. It is a researched-based model of sheltered instruction, comprised of eight components and 30 features that focuses on teaching academic content and language simultaneously.

[4] The LATIC model is a "student-centered, problem-based model for learning" (http://www.idecorp.com/) designed by Innovative Designs for Education.

workshops as teachers, while Vivian is the individual who provides this training to all teachers (new and those who are assigned as a refreshed) in the district. According to them, SIOP® provides a common language, one with a culturally responsive focus, for individuals at all levels of the system to use when talking about teaching and learning.

Since teachers are expected to incorporate strategies learned from "The Skillful Teacher" and the SIOP® and LATIC trainings into their instruction, many culturally responsive instructional practices are embedded in the district's teacher practice rubric, as well as in the supervision and evaluation process. Ben explains:

> It is our expectation is that every teacher should be able to be culturally responsive. They all went through SIOP. They can take a refresher. I just assigned a teacher to SIOP last summer. I think SIOP could teach [him] how to be more empathetic and understanding, and [give him] the ability to differentiate for all kids. It's just best practices.

Hence, at Central Bridge, ESOL-focused PD is a fundamental part of culturally responsive teaching. Teachers are given multiple pathways to grow as professionals and to ensure that their teaching meets the unique needs of English language learners.

At the classroom (micro-) level, Anita has described these PD workshops and trainings as "an unbelievable powerful experience," and she has brought the strategies she has learned back into the classroom. In reviewing classroom video-recordings of her instruction and observational notes, there are several strategies from the aforementioned PD workshops and trainings that Anita has integrated into her instruction. First, in terms of *The Skillful Teacher*, she has an effective management system in place; she demonstrates a strong pedagogical knowledge (how to teach) as well as a knowledge of her individual students; she has the knowledge of the content and its linguistic demands and knows what kinds of supports to give her students so that they can access the content. Next, in each of the three observed lessons, Anita has incorporated the eight SIOP® components into her instruction: lesson preparation, building background, comprehensible input, strategies, interaction, practice and application, lesson delivery, and review and assessment. These components are not in place as part of a checklist of her own behaviors, but rather to support student learning, to meet the students where they currently are, and to help them reach the next level. Finally, in her classes, Anita provides student-centered and hands-on instruction, allowing students to actively participate and interact with her and their peers, following the LATIC model. Moreover, she incorporates technology in purposeful ways to differentiate and support student learning when appropriate. In Anita's words, "I would say all of this professional development really allows us to offer a wide range students very valuable, very relevant learning experiences."

Table 6.2 includes an excerpt from a transcript of Anita's class, and it shows how she has integrated the theories she learned in the PD training into her instruction, and thus employing a culturally responsive teaching approach:

After viewing the video recording of this lesson, Anita wrote a reflection, describing what she saw and how she interpreted both her actions and those of her students. First, Anita noted that she paired students purposefully to support their learning. For example, a new student had just begun this class, and she paired him up with another student who "is patient, kind, and strong in English and science. [Student] seemed like the right match." These students worked well together, speaking both in English and in Spanish, and earned a perfect score on the lab. Secondly, Anita had noticed in previous lessons that students were not interacting with one another as much as she would have liked in English. Therefore, she prepared a set of sentence frames to scaffold this interaction and encourage students to use English as they worked in pairs. She noted "despite the fact that most of the students did not use the sentence frames, there was more English used in their interaction during independent work than in the last lesson we recorded." Finally, Anita highlighted the many times she tried to activate students' background knowledge and make connections to their prior learning experiences. For example, with the "do now" activity, she noted that it was "an important activity for the four students who were not in the class during unit 2, since they would not have had this information as background knowledge." To conclude, it is evident that Anita has a working

Table 6.2 Excerpt of classroom instruction

T: S1? Can you read the do now? S1: What do these diagrams represent? Why are they important? How are they similar, different?	-"Do now" activity is a review of previously learned vocabulary and concepts; activates prior knowledge on topic.
T: Okay, so these two diagrams represent something we've already learned about and you can look in your notes. Okay? These are the structural formulas of two things. Wait wait wait—yes? (Some students give answers.)	-Uses wait time to provide the opportunity for all students to engage with the activity.
T: I'm hearing really good words and I can see S2 has a really good idea. Okay but before we share, go back in your notes and think. What am I showing you? What is the difference between these two things? How are they similar? What do organisms use it for? Okay? So, S3, what are you thinking? S3: The one on the left is a molecule of glucose. T: The one on the left is a molecule of glucose. Do you agree with him? Ss: Yes	-Encourages students to use higher order thinking skills, comparing, and contrasting.
(Teacher takes students through this same procedure for identifying starches, then moves on to a review of diffusion.) T: Okay (writes on white board). I want you to think about this because we are talking about how things can go through the cell membrane, right?	-Classroom procedures have been established and students know the routine.
Remember? Diffusion, yes? Okay. Okay so now that we understand the difference and we now remember the difference between glucose and starch, we're going to work on a lab. The lab has three parts (holds up three fingers). Today we're doing part one and it's on indicators. So who can tell me what is an indicator? S4: A substance that indicates that there's (unclear) T: A substance that indicates the presence of another substance. How? How do you know?	-Continues to connect previously learned concepts and vocabulary in preparation for this lesson's lab.

(*continued*)

Table 6.2 (continued)

Ss: Changes color	-Asks students to provide her with evidence.
T: By changing color. Okay. So let's read our objectives (projects objectives onto white board) and then we're going to start working with indicators. S4 can you read the content objective please?	
S4: Students will work with two substances to discover if they are indicators for starch or glucose.	-Explicit review of objectives with students.
T: Okay so our lab when we start working on the lab we are going to work with indicators for which two substances?	-Checks for understanding.
S5: Glucose and starch	
S6: Starch and glucose	
T: We're going to find out which liquid, which chemical, can we use to see if a solution has starch and which chemical we can use to see if a solution has glucose. Okay? And the language objective, S7 can you read the first bullet?	
S7: Students will test for the presence of glucose and starch in a solution by following written steps	
T: Okay, so you're going to very carefully read the lab and with your partner follow the steps. Okay? And make sure to use checkmarks (gestures as if making a checkmark) to know where you are in your lab procedures. And what's the second bullet S8?	-Classroom management system in place to support learning.
S8: Communicate clearly with a lab partner to successfully complete the lab.	-Student-centered approach: students will work in pairs so that they have the opportunity to interact with one another and engage with the content and language in authentic ways.
T: Okay so let's do a demo on indicators (switches projected screen). Okay what's an indicator? A substance that changes . . . ?	
Students: color	
T: color when it reacts with	
S: something else	
T: Something else, right? So I spent a lot of time grinding cabbage the other day (holds up a cabbage). Have you ever had cabbage?	-Models what the students will be doing in the lab.

S: Yeah	
S: No.	- Pedagogical knowledge
T: No? Have you ever had coleslaw?	
S: What?	
T: That salad they serve you with the hamburger sometimes?	
S: Coleslaw	
S: Yes! Coleslaw.	
T: Okay coleslaw, yes? Okay so this one is red cabbage, coleslaw is made with white cabbage. Okay? So if you grind this in a blender, you get this (holds up graduated cylinder with cabbage juice). What color is it?	
Students: Purple	
T: Yeah it's like very dark purple, right? Okay so the red cabbage juice is an indicator and scientists have worked with it and they found out it changes colors depending on whether a substance is an acid or not an acid. Now I know we didn't talk about acids in the class . . . If I used red cabbage juice (holds up cylinder) with something that's an acid, it will turn this color (points to white board). What color is that?	-Makes a connection to students' prior experiences with cabbage. -Builds background for those students who are not familiar with it.
Students: Red.	
T: Red. Okay? If I want to test something different, and it turns green (points to white board), then it's not an acid. It's something we call a base. And a base is the opposite of an acid. Okay?	-Connects students' experience and background with cabbage to the scientific concept they will study in class.
(Teacher sets students up with partners and reviews directions and sentence frames to encourage students to interact with their partners in English)	-Reviews directions to make sure students understand what is expected of them.

classroom management system in place, she has knowledge of both the content and its linguistic demands, and she knows her students and how to reach them. As such, she is culturally responsive in her practice.

In interviewing some of Anita's students after this class, several identified doing labs and experiments as both challenging, yet at the same time, what they like most about learning science. The following is an excerpt from this interview:

Interviewer—Could you tell me about a time when you had a difficult experience with science?
S1—Doing the experiments. Following the directions.
S2—Doing the experiments is the most difficult thing in science. Cause you have to follow the directions and if you do not, everything is wrong.
Interviewer—So if you don't follow the directions, it doesn't matter what you do?
Ss—Yeah (laughing)
Interviewer—Are there things your teachers do to help you when science is difficult?
S1—Yeah. She review the directions.
S4—Tell you to follow the directions.
S3—Explain the difficult words.
Interviewer—What do you like most about learning Science?
S2—New things.
S1—Experiments.
Interviewer– You say experiments but you said they were hard, but you like them? Can you explain that?
S1—Cause they are new ones. You use the microscope.
S4—You learn more.
S1—You get to explore more things.
Interviewer—Do you like Science?
Students—Yes!

Their responses to the questions demonstrate that the aforementioned practices that Anita has integrated into her instruction have influenced student learning in positive ways. Taking the time to explain directions and explicitly teaching key vocabulary are instructional practices that have helped these students overcome the challenges of this content area and give them meaningful learning experiences. Furthermore, these instructional practices are evidence to support the notion that by implementing what she learned in her PD experiences, Anita is poised to meet the unique needs of her students, to build upon their strengths and previous experiences, and to challenge them within this sheltered context in preparation for what they will face the following year in a mainstream class.

Differentiation

While they have made a concerted effort at Central Bridge to initially shelter students as they are learning the language and then gradually release them into the mainstream, science classes at Central Bridge are still quite diverse in terms of age, proficiency, ability, and levels of former education. Yet, because of the ESOL-focused PD, collaboration within and across departments, quest for equity and accessibility, and teacher dispositions, teachers feel that they are able to serve a very diverse group of students in the same classroom during a 55-minute period. Anita noted "I am not saying every single day I am able to differentiate everything, but we're able to reach all of our students where they are, and challenge them to move up to that next level of proficiency, in a very challenging environment... you know i+1.[5]"

At both the building and district level, there is an expectation that teachers will be able to differentiate, which is to say to take what they have learned in the *Skillful Teacher,* SIOP®, and LATIC PD workshops and apply them in the classroom. The Central Bridge Teacher Practice Rubric communicates the expectations of practice, and it is used in the supervision and evaluation process. There are four key organizers to this rubric which include: knowing how to teach and how to teach it to each individual,

[5] i+1 refers to the level of comprehensible input when students are learning a new language. The "i" represents the current competence of the learner, while the "1" is the next level a little beyond the current level. According to Krashen (1982), students acquire language in a low anxiety environment, when they do not have to speak until they are ready to, when input is "interesting, relevant, and not grammatically sequenced," and when error correction is minimal.

knowing the students and how each one learns, managing and monitoring progress for student success, and creating a climate and culture for student learning. As Ben indicated in his interview, teachers need to be able to differentiate and to meet the needs of all of their students in order to be deemed "highly effective." He stated "meeting the needs of all learners and appreciating the diversity of our district and understanding English language learners . . . [it] is a priority. It's big and it's who we are."

Paul also shared his perspectives on differentiation, specifically in the science classroom. In his interview, he called attention to the section of the rubric that addresses the following criteria: teachers engage all students in meaningful and relevant learning by differentiating instruction through varied content, processes, and products in response to students' prior knowledge and skills. In his view, one effective way to meet this end is through problem-based learning where teachers open up a question and the kids come at it with what they have. For him, it is important for teachers of ELLs to "get [them] in a laboratory with a problem and it's just sitting there, hands-on, in front of them, some skills emerge that you would not have expected because it's hard for them to put that into language that we typically use as teachers to assess kids."

At the classroom level, both Anita and Lisa view the ability to differentiate as central to their practice. In one of her observed lessons, Anita incorporated a modified version of a flipped classroom where half of the students listened to a video lecture while completing scaffolded notes, while the other half worked with her to apply key concepts from the power point presentation they viewed in the previous class. The students were grouped homogenously which allowed her to differentiate the video presentation and the activity. The group watching the video presentation was at a lower level than the second group, and they took advantage of the opportunity to watch the video at their own pace, to pause as they filled in the notes, and to rewind as necessary. Anita was able to give the second group more individualized attention, assistance, and feedback as they worked collaboratively to complete the activity related to the video they had watched in the previous class. In her reflection on the lesson, Anita noted the ways in which she differentiated this instruction, as well as her reasoning behind her instructional decisions. Additionally, she noticed how much her students had grown as independent learners. She wrote:

> One of our goals is to work toward students working independently and taking responsibility for their own learning. After watching this session, I believe my class has made huge strides in that direction. Overall, there was a sense of order and independence that made me appreciate how hard the students were working at being responsible learners.

Similarly, Lisa acknowledged that in her mainstream Living Environment class, there is a veritable need for differentiated instruction. She looks at each student as an individual and designs a variety of instructional activities to support each one as they are learning both the language and content simultaneously. In an observed lesson on the digestive system, the ways in which she differentiated to meet the varied needs of her students included: providing scaffolded notes and visuals, using Brain Pop for vocabulary reinforcement, providing demonstrations of the digestion of food to add to visual experience, and allowing students their choice in how they would complete activities (individual, with a partner, or as a small group with her assistance). While the native-English speaking students chose to work either independently or with a partner, the ELLs chose to work with her. This instructional decision allowed Lisa to be able to provide them with individualized assistance with both the content and the language. She found that these supports were effective for the ELLs in this mainstream, Regents-level class as each was able to complete the lab on digestion successfully. One of her students indicated that this strategy has been helpful to him because the vocabulary on labs is challenging. He said "she explains things to me. She goes over the notes. She tells me to read carefully, so I read everything twice. And if I don't understand, she will give me the help I need."

Discussion

Anita and Lisa are examples of two teachers who have embraced the idea of culturally responsive teaching at Central Bridge High School, and student learning has been positively influenced as a result of their pedagogy. The district and building leadership identified both teachers as having a disposition that is well-suited for working with ELLs, and for both of Anita and Lisa, the student is the focus in their planning and instruction. They have employed many of the ideas from the PD that was provided by the district, and as such, their classrooms are culturally responsive to their ELL students' needs. Moreover, both teachers have been able to meet students at their current level by differentiating their instruction and providing all students with meaningful learning experiences. These teachers have an ability to reach ELLs and engage them in authentic ways with both the content and the language. Finally, their empathy and quest for equity also promote a culturally responsive approach inside and outside the classroom. They view their students as emergent bilinguals, with many resources and prior experiences that play an important role in their learning. Anita explained "my mission is to make sure I do everything I can to have a student feel empowered to live his life at his fullest in English." Lisa added "they are my heroes." For these teachers, learning is not equated to a test score, but rather to meaningful experiences that will prepare them to be full participants in our society.

Implications for Teacher Preparation

The ESOL-focused PD provided by the Central Bridge School District to its teachers can be used as a model for pre-service teacher preparation and in-service teacher PD alike. This district acknowledges that ELLs come to the classroom with linguistic diversity and cultural differences, and both are viewed as resources and assets for students in the science classroom. The PD they offer teachers positions them to meet all students at their individual current level, to use their previous experiences (both in and out of the classroom) as a bridge to new learning, and to differentiate their instruction to reach all students in the classroom. At Central Bridge, there is an equal emphasis placed on both what to teach and how to teach it. As such, this model can be viewed as a way of closing the achievement gap in science among ELLs and NES students.

Science teachers who have not had prior coursework or training in TESOL methods can meet the needs of linguistically and culturally diverse students when they employ a culturally responsive approach to their teaching. While the SIOP® model has been criticized by some as it is considered to be a "one-size-fits-all" approach to teaching ELLs in mainstream classes, it can serve as an entry point for these teachers as they consider not only the content they teach, but also the linguistic demands that are associated with this content. Instead of focusing solely on the 8 features and 30 components of the SIOP® model in their practice, teachers should also consider how a culturally responsive and student-centered approach to teaching would provide ELLs with meaningful and authentic ways to engage with the content and language. Finally, they should be mindful of the need to differentiate in order to reach all students, regardless of their native language and/or prior experience with science.

References

Bronfenbrenner, U. (1977). Toward an experimental ecology of human development. *American Psychologist, 32*(7), 513–531.

Committee on Multicultural Education (2002). *Educators' preparation for cultural and linguistic diversity: A call to action.* Retrieved from: http://www.buildinitiative.org/portals/0/uploads/documents/resource-center/diversityand-equity-toolkit/culturallinguistic.pdf.

Creswell, J. W. (2013). *Qualitative inquiry and research design: Choosing among five approaches* (3rd ed.). Thousand Oaks, CA: Sage.

de Jong, E. J., & Harper, C. A. (2005). Preparing mainstream teachers for English language learners: Is being a good teacher good enough?. *Teacher Education Quarterly, 32*(2), 101–124.

Dewey, J. (1938). *Experience and education.* New York: Macmillan.

Echevarria, J., Vogt, M., & Short, D. (2013). *Making content comprehensible for English learners: The SIOP® model.* Upper Saddle River, NJ: Pearson.

Freeman, D., & Freeman, Y. (1988). *Sheltered English instruction.* Retrieved from http://www.ericdigests.org/pre-9210/english.htm.

Freire, P. (1970). *Pedagogy of the oppressed.* London: Harmondsworth: Penguin.

Gebhard, M., & Willett, J. (2008). Social to academic: University-school district partnership helps teachers broaden students' language skills. *Journal of Staff Development, 29*(1), 41–45.

González, N., Moll, L. C., & Amanti, C. (2005). *Funds of knowledge: Theorizing practice in households, communities, and classrooms.* Mahwah, N. J.: L. Erlbaum Associates.

Hansen-Thomas, H. (2008). Sheltered instruction: Best practices for ELLs in the mainstream. *Kappa Delta Pi Record, 44*(4), 165–169.

Jiménez, R., David, S., Pacheco, M., Risko, V., Pray, L., Fagan, K., & Gonzales, M. (2015). Supporting teachers of English learners by leveraging students' linguistic strengths. *Reading Teacher, 68*(6), 406–412. doi: 10.1002/trtr.1289.

Knight, S. L., & Wiseman, D. L. (2006). Lessons learned from a research synthesis on the effects of teachers' professional development on culturally diverse students. In K. Tellez & H. C. Waxman (Eds.), *Preparing quality educators for English language learners: Research, policies, and practices* (pp. 71–98). Mahwah, NJ: Lawrence Erlbaum Associates.

Krashen, S. (1982). *Principles and practice in second language acquisition*. Oxford, U.K.: Pergamon Press.

Lave, J., & Wenger, E. (1991). *Situated learning: Legitimate peripheral participation*. Cambridge, UK: Cambridge University Press.

McIntyre, E., Kyle, D., Chen, C., Muñoz, M., & Beldon, S. (2010). Teacher learning and ELL reading achievement in sheltered instruction classrooms: Linking professional development to student development. *Literacy Research & Instruction, 49*(4), 334–351. doi: 10.1080/19388070903229412.

Miles, M., Huberman, M., & Saldana, J. (2013). *Qualitative data analysis: A methods sourcebook*. Thousand Oaks, CA: Sage.

Murray, D. E. (2009). The ecology of leadership in English language education. In M. Christison & D. E. Murray (Eds.), *Leadership in English language education: Theoretical foundations and practical skills for changing times* (pp. 13–26). New York: Routeledge.

National Center for Education Statistics. (2009). *The nation's report card*. Retrieved from: http://nces.ed.gov/nationsreportcard/pdf/main2009/2011451.pdf.

NGSS Lead States (2013). Next Generation Science Standards, Case Study 4: English Language Learners and the Next Generation Science Standards. Retrieved from: http://www.nextgenscience.org/sites/default/files/%284%29%20Case%20Study%20ELL%206-14-13.pdf.

Nieto, S. (2000). *Affirming diversity: The sociopolitical context of multicultural education* (3rd *ed.*). Reading, Massachusetts: Longman.

Oliver, B., & Oliver, E. (2013). Culturally responsive teaching: How much more data do we need?. *Journal of International Education & Business.*, *4*(1), 5–25.

Peercy, M. M. (2011). Preparing English language learners for the mainstream: Academic language and literacy practices in two junior high ESL classrooms. *Reading & Writing Quarterly, 27*(4), 324–362. doi: 10.1080/10573569.2011.59105.

Piaget, J. (1936). *Origins of intelligence in the child.* London: Routledge & Kegan Paul.

Portin, B. (2004). The roles that principals play. *Educational Leadership, 61*(7), 14.

Ramirez, P. C., & Jimenez-Silva, M. (2014). Secondary English language learners: Strengthening their literacy skills through culturally responsive teaching. *Kappa Delta Pi Record, 50*(2), 65–69. doi: 10.1080/00228958.2014.900846.

Téllez, K., & Waxman, H. C. (2006). Preparing quality teachers for English language learners: An overview of the critical issues. In K. Tellez & H. C. Waxman (Eds.), *Preparing quality educators for English language learners: Research, policies, and practices* (pp. 1–22). Mahwah, NJ: Lawrence Erlbaum Associates.

Theoharis, G., & O'Toole, J. (2011). Leading inclusive ELL: Social justice leadership for English language learners. *Educational Administration Quarterly, 47*, 647–688.

Verma, G., Pepper, J., & Martin-Hansen, L. (2008). Effectively communicating with English language learners using sheltered instruction. Science Scope, 31(13), 56–59.

Vygotsky, L. S. (1978). Mind in society: The development of higher psychological processes. Cambridge, MA: Harvard University Press.

Vygotsky, L. S. (1986). *Thought and language.* Cambridge, MA: MIT Press.

Walqui, A., & Van Lier, L. (2010). *Scaffolding the academic success of adolescent English language learners.* San Francisco: WestEd.

Wardle, F. (1996). Proposal: An anti-bias and ecological model for multicultural education. *Childhood Education, 72*(3), 152–156.

Wepner, S. B. (2011). *Changing suburbs, changing students.* Thousand Oaks, CA: Sage.

Wilcox, K. C., Lawson, H. A., & Angelis, J. I. (2015). Classroom, school and district impacts on minority student literacy achievement. *Teachers College Record, 117*(10). Retrieved from http://www.tcrecord.org/content.asp?contentid=18049.

Yin, R. K. (2014). *Case study research: Design and methods* (5th ed.). Thousand Oaks, CA: Sage.

Gretchen Oliver is a doctoral candidate in the Department of Educational Theory and Practice at the State University of New York at Albany. She is the coordinator for a U.S. Department of Education grant-funded research and professional development project, the Technology-Enhanced Multimodal

Observation Protocol Project. In this capacity, she has provided professional development to secondary math, science, and ESOL professionals in an effort to help them augment academic English through the content areas, as well as math and science content through academic English. In addition, she teaches Teaching English to Speakers of Other Languages method courses to pre-service teacher candidates and in-service content/classroom teachers.

7

Writing Their Way into Science: Middle School English Language Learners as Next Generation Scientists

Judy Sharkey and Tina Proulx

How can a middle school science teacher help her English language learners (ELLs) develop the literacy skills necessary for the science knowledge and thinking practices promoted by the Next Generation Science Standards (NGSS) and the Common Core State Standards (CCSS)? In this chapter we share our pursuit of this question, combining a writing process approach to content development with a commitment to recognizing and valuing the rich cultural and linguistic identities that these students bring to our schools and communities. The content focus of this chapter is a ten-week unit on the hydrosphere, one of the Earth's four subsystems, designed and taught to a diverse group of English learners in a seventh grade sheltered science classroom.

J. Sharkey (✉)
Department of Education, University of New Hampshire, Durham, NH, USA
e-mail: Judy.Sharkey@unh.edu

T. Proulx
Manchester School District, Henry J. McLaughlin Middle School, Manchester NH, USA
e-mail: tproulx@mansd.org

© The Author(s) 2017

L.C. de Oliveira, K. Campbell Wilcox (eds.), *Teaching Science to English Language Learners*, DOI 10.1007/978-3-319-53594-4_7

We begin with a brief introduction to the nature of our collaboration and the key tenets of our approaches to teaching multilingual learners in US public schools. Next we describe our local context, leading into an overview of the unit: its objectives, standards, a curriculum framework, and key instructional activities. We then introduce the voices and learning of four focal students to show how they were developing as young scientists. The collection of these students' literacy and content development processes and products serve as rich examples of what students, even at beginning levels of English language proficiency, can achieve in science. We conclude with insights learned from our inquiry and implications for next steps.

A Practitioner and Researcher Collaboration

As a public school teacher/curriculum consultant and a university-based teacher educator/researcher, we have been collaborating for close to a decade on a range of endeavors from designing and providing professional development on culturally and linguistically responsive pedagogies, mentoring/supervising prospective English to speakers of other languages (ESOL) teachers, and supporting community partners that serve our immigrant/refugee populations. We see our collaborations as mutually beneficial; the differing perspectives we bring to our shared inquiries strengthen our abilities and effectiveness as classroom teachers, teacher educators, and researchers. Visiting Tina's classes on a regular basis brings a much needed reality check to Judy's graduate ESOL methodology and curriculum course content; seeing and hearing her practice through observation notes followed up by discussion forces Tina to articulate and challenge her pedagogical rationale and assumptions. At the time of this project, Tina was working on a district level curriculum development team advocating for richer, more rigorous content for English learners, and Judy was working with mainstream educators who were struggling with ways to connect STEM content to their ELLs' cultural and linguistic identities and realities. The ten-week science unit allowed us to ground our current challenges in a classroom context that would resonate with the teachers in our projects.

Judy visited the class 14 times over three months, starting two weeks before the hydrosphere unit in order to get to know the students and the rhythm and flow of the class period. She kept written notes and we would talk for 20–30 minutes after the class. We started a shared notebook on Google docs where Judy would type up her observation notes—a combination of what happened, bits of dialogue when possible, and interjecting questions. Sometimes the questions were about a particular student (e.g., "What's up with Alejandro today?"); sometimes they were about an instructional practice (e.g., "How do you use cognates with the Spanish speakers?"). Tina would then read and respond to notes, often adding wonderings and questions that took us in new directions or indicated that we needed to clarify our positions/thoughts. On days when Judy wasn't there, Tina used her laptop computer to audio-record the classroom talk. We also met several times at Judy's office on campus to discuss challenges, issues, and look at the work students were producing.

Teach Students not Content, but Don't Forget about the Content!

Our work is informed by critical pedagogies and social justice education. Paulo Freire's (1988) work in humanizing education and positioning students as partners in creating curriculum and inquiries and Jim Cummins' assertion that "human relationships are at the heart of schooling" (2001, p. 1) are integral to our philosophies. As Cummins (2001) eloquently states, "The interactions that take place between students and teachers and among students are more central to student success than any method for teaching literacy, or science or math" (ibid.). Developing healthy relationships with students begins with affirming their identities, their lives, and experiences outside of schools. Research has shown that a learner's first language (L1) can be an important resource in second language (L2) learning (see e.g., De Jong and Harper 2005) but equally important is recognizing students' funds of knowledge, the rich array of knowledge students bring with them from their families' histories and homes and that fosters engagement and learning in schools (Moll et al. 1992). In line with Freire's notion of students as subjects of their own learning, Joves et al. (2015) argue for teachers' incorporating into their classrooms "funds of identities," the new skills and knowledge that students build that may or not be connected to their families' funds of knowledge. In addition to the humanistic reasons for affirming students' identities, research has shown that healthy student-teacher relationships are connected to academic achievement, particularly for immigrant adolescents (Suárez-Orozco et al. 2008).

Teaching from a social justice perspective means being concerned with issues of equity, access, and representation (Nieto 2010). From a curriculum standpoint, this means ensuring that students' lives and interests are visible in the curriculum, something that unfortunately is not the case for many of our culturally, linguistically, and racially diverse student populations (Cochran-Smith 2004) and something that has become more challenging in this era of standardization and accountability (Comber, 2016; Compton-Lilly and Stewart 2013). In the science classroom this means attending to who is represented in pictures/visuals of scientists and including examples of innovations and breakthroughs from all different types of contexts. Teaching students to reach mandated standards is not antithetical to social justice, but we must do so in critical ways, making sure that students' lives and aspirations are part of the content and inviting students to question what is included/excluded in the standards and why (Kumashiro 2009).

Lastly, teaching from a social justice perspective also means providing academically and cognitively challenging content that opens up career pathways and trajectories students might not have previously imagined. For teachers of English learners this also means supporting the development of academic oral language and discourses (Gee 2005). For example, once students understand that "potable water" means drinking water, teachers should use and expect students to use "potable." It also means integrating metalinguistic skills into the curriculum so students can use their knowledge of language subsystems in understanding and expressing their growing content knowledge (Palincsar and Schleppegrell 2014). An example in the science classroom is when we draw attention to the relationship between "evaporate" and "evaporation," thereby helping students to notice nominalization, a linguistic aspect integral to acquiring science discourses (Gibbons 2015). Just as we strive to see students as who they are when they come to our classrooms, we must also see them as who they could and/or will be and then let those trajectories inform our content and language instruction. The subtitle of our chapter captures this sentiment. Shifting from teaching NGSS to next generation *scientists* reflects our commitment to helping students add "scientist" to their funds of identities and aspirations.

Contexts: City, School, and Classroom

Manchester is a small New England city (pop. @110,000) whose immigrant identity can be traced back to the mid-nineteenth century. There is a growing, vibrant Latino population, and it is a refugee resettlement city. In the last seven years, the city has resettled close to 1600 refugees with the largest groups from the Democratic Republic of the Congo, Sudan, Somalia, Nepal (the Nepali-Bhutanese), and Iraq. There are over 80 different languages in the district with Spanish, Arabic, and Nepali representing 90% of those languages. Similar to patterns seen in urban centers across the nation, the schools with the highest concentration of culturally, linguistically, and racially diverse students are also the ones with the highest number of low-income students. Tina teaches at McLaughlin Middle School (MMS) where approximately 10% of the 748 students receive EL services and over half of all students (53.6%) are eligible for free or reduced price lunch. MMS uses a tiered approach to EL instruction. Newcomers are placed in age/grade appropriate self-contained classrooms that use sheltered instruction to deliver mainstream curriculum. The goal is to have students exit the self-contained program after 12–18 months. The second tier of the EL program mainstreams students into all content classes but also provides them with one literacy support class per day. Tina teaches six periods a day, including three science classes (grade six, seven, and eight).

When the school year started in September, there were seven students in Tina's seventh grade science class. By the middle of February the class had doubled to fourteen. This group consisted of eleven boys and three girls, ages 12–14. The nine Spanish-speakers were from the Dominican Republic; El

Salvador; Honduras; Mexico; and Puerto Rico. The five remaining students were from Somalia (2), Iraq (1), Burma (via Malaysia) (1), and Sudan (1). At the time of this unit, 50% of the class were at WIDA[1] level (1): entering; the other 50% were emerging (2) or developing (3). The class met five days a week for a 46-minute period at the end of the day.

The challenge with students arriving at any moment in the school year means that they must quickly be socialized into the practices of the classroom and in a welcoming way. Even though Tina had devoted the early part of September to introducing the scientific inquiry process, it was new to the seven students who had arrived between December and February. Also, because students are progressing in the language/content development at different rates based on a number of factors, including previous schooling, first language (L1), and physical/mental health issues, stress related to negotiating multiple new contexts, etc., teachers must constantly be thinking about how to make content concepts accessible yet challenging for all students. The instructional practices that Tina uses to address these challenges include establishing predictable, known classroom routines, assigning L1 language buddies when possible, and providing lots of multisensory resources and scaffolding (Peregoy and Boyle 2013). These instructional strategies are explained in more detail within the description of the unit.

Curriculum Framework: Writing Science and Doing Science

According to the NGSS (2013),

> [L]iteracy skills are critical to building knowledge in science... Likewise, writing and presenting information orally are key means for students to assert and defend claims in science, demonstrate what they know about a concept, and convey what they have experienced, imagined, thought, and learned (p. 1).

[1] The state belongs to the WIDA Consortium and uses WIDA products for placement, standards, and annual assessment of academic language growth.

ESOL teachers are trained to attend to learners' literacy development across the content areas, but in addressing the literacy aspect of science teaching, Tina drew on her history as a language arts teacher and active participant in the National Writing Project (NWP). Her curriculum reflects the NWP philosophy: "Writing is essential to communication, learning, and citizenship... Writing helps us convey ideas, solve problems, and understand our changing world. Writing is a bridge to the future" (National Writing Project n.d.). Details of how Tina incorporated writing into her science curriculum are provided in the description of the unit.

The reading, writing, and talking about science does not and cannot replace the doing of science both in content/concept development and in students' overall understanding of what it means to be a scientist. As the National Resource Council's (NRC) framework for K-12 science education asserts:

> The actual doing of science or engineering can also pique students' curiosity, capture their interest, and motivate their continued study; the insights thus gained help them recognize that the work of scientists and engineers is a creative endeavor—one that has deeply affected the world they live in. Students may then recognize that science and engineering can contribute to meeting many of the major challenges that confront society today, such as generating sufficient energy, preventing and treating disease, maintaining supplies of fresh water and food, and addressing climate change. (NRC 2012, pp. 42–43)

Curriculum Unit: Inquiries into the Hydrosphere

An Understanding by Design (UbD) (Wiggins and McTighe 2006) approach guided the design of the ten-week unit that Tina taught beginning in February 2016. Consistent with UbD, Tina used a recursive process that entailed identifying and articulating the key understandings, formulating an essential question, designing appropriate performance tasks, determining evidence of learning, identifying appropriate standards, and sequencing learning activities. The culminating activity required the students to work in teams to design, test, and evaluate a prototype for a water filtration system.

The hydrosphere unit. Essential questions: How does learning about the water cycle help us understand Earth's systems? How do humans impact the Earth's water system?

Goals and objectives (students will be able to): a) explain and discuss the cycling of water through Earth's systems driven by energy from the sun and the force of gravity (the water cycle); b) recount and explain as to why scientists study the water cycle; c) explore scientists who do the research to understand how human activity impacts natural resources; d) research the human population per-capita consumption of water; e) write about the impact of water consumption on Earth's systems; f) construct a model for a phenomenon or a water-related problem.

Evidence of learning: water cycle diagram; water cycle model; letters to dialogue journal partner; daily quick writes; notes; materials test conclusion; water filter models; class discussions; student conversations; and reflections of learning (oral and written)

Key vocabulary: evaporation, condensation, precipitation, transpiration, infiltration, groundwater, runoff, water cycle, water vapor, erosion, deposition, agriculture, industry, potable, Earth's systems (hydrosphere, atmosphere, geosphere, biosphere), crisis, system, cycle, filter, prototype, clarity, speed

Standards[2]. The NGSS, CCSS, and WIDA were the sources of the standards. Here we share a few of the key standards addressed.

CCSS.ELA-LITERACY.WHST.6–8.10. Write routinely over extended time frames (time for reflection and revision) and shorter time frames (a single sitting or a day or two) for a range of discipline-specific tasks, purposes, and audiences.

[2] Manchester school district uses the NGSS and WIDA standards. They used the CCSS to write their own academic standards. In this chapter, we list the CCSS that were the original source for Tina's unit.

NGSS MS-ESS2–4. (Middle school-earth systems). Develop a model to describe the cycling of water through Earth's systems driven by energy from the sun and the force of gravity.

NGSS MS-ETS1–1. (Middle school engineering design). Define the criteria and constraints of a design problem with sufficient precision to ensure a successful solution, taking into account relevant scientific principles and potential impacts on people and the natural environment that may limit possible solutions.

NGSS MS-ETS1–3. (Middle school engineering design). Analyze data from tests to determine similarities and differences among several design solutions to identify the best characteristics of each that can be combined into a new solution to better meet the criteria for success.

WIDA English language development standard 4. ELLs communicate information, ideas, and concepts necessary for academic success in the content area of Science.

WIDA created Can Do descriptors to highlight what students can do with language. These are organized into four categories of communicative purposes: recount, explain, argue, and discuss and apply to the four modalities: listening, speaking, reading, and writing. In the hydrosphere unit, examples for Can Do statements for explain were:

Listening: Identify functions of content-related topics based on short oral statements reinforced visually; carry out a series of oral directions to construct scientific models

Speaking: Demonstrate how to conduct experiments, engage in processes, or solve problems with supports

Reading: Match content-related objects, pictures, or media to words and phrases; compare and contrast information from experiments, simulations, videos, or multimedia sources with that from reading text on the same topics

Writing: Indicate relationships by drawing and labeling content-related pictures on familiar topics; describe processes or cycles by labeling diagrams and graphs; produce descriptive paragraphs around a central idea

Tina broke down the larger unit into three phases, each with its own guiding question. Weeks 1–3: What is the water cycle and how does it work? Weeks 4–6: Who studies the water cycle and its potential impacts on Earth's systems? Weeks 7–10: What are some of the most pressing water problems? What are some design solutions to a water-related problem?

Teaching the Unit: Writing Science and Leveraging Cultural and Linguistic Resources

Tina used three instructional writing techniques to support student learning: daily quick writes; ongoing science journals; and weekly dialogue journal partners. The daily quick writes are the posted questions that start each lesson. Tina drafts questions when mapping out her curriculum but then adjusts them each day to reflect what happened in the previous class. As students enter the classroom, they immediately take out their science journals, copy the question, and begin writing a response. In addition to helping establish a predictable routine, the questions serve as a type of posted lesson objective that helps ELLs focus their learning. As Echeverría, Vogt, and Short (2014) recommend, they should reflect what students "will learn or do, . . . be stated simply, . . . and tied to specific grade-level content standards" (p. 29). Here are two examples from the unit:

> *We are going to continue talking about who studies the water cycle. Why do you think some scientists study the water cycle?* (2/2/16)
>
> *We will continue to talk about how humans affect the hydrological cycle of the earth. What are some ways/reasons humans use water every day?* (3/1/16)

As students began writing, Tina would circulate to make sure each student understood the question. Some needed Tina to orally re-phrase or find a translation but they would write the question as it was posted. Each day a different student read the question aloud and several students read their responses.

The daily quick writes also helped organize the students' science journals where they recorded notes on discussions and experiments, drew models, could use L1 and illustrations as necessary, and reflected on their learning. The examples of the quick writes above also worked as class brainstorms. As students shared their thoughts regarding ways of and reasons for daily water use, Tina created a large list on the board. Students then wrote the bulleted information in their journals.

Another key writing practice was the weekly dialogue journal, an activity that adds a more authentic and communicative purpose to students' writing about science, and provides opportunities for ELLs to interact with native-English-speaking peers in other classrooms. Here is the letter that Tina wrote to her MMS colleagues in February:

Dear Science Teaching Colleagues,
I am looking for 14 students to be reading/writing response partners for my 7th grade science class during the month of March. We have just finished studying the water cycle and now we are looking at the human impacts piece. I was just thinking that if each of my students had a pen pal to write to about their science inquiries, the experience would support my students' language and content learning while also providing a purpose for engaging in writing activities. How the writing partnership would work is I would like one meeting time (face to face) during next week so the writing partners could get to know each other. Then my class would write our first letters to our partners on Friday March 4th. The writing partner would respond on Tuesday March 8th. Then, the cycle continues with my students writing on Fridays and your students responding on Tuesdays. The writing dates would be as follows: My students: Fridays—March 4, 11, 18, 25. Your students: Tuesdays—March 8, 15, 22, and 29. My students will be watching videos the first week, researching information the second week and then, in the third and 4th week, engineering a design solution to a water problem they identify.
Thank you for considering the idea,
Tina Proulx, ELL Science & Language Arts Teacher

Complementing Tina's commitment to writing, Judy wanted to see how the curriculum could reflect students' identities and global perspectives. Her mainstream teachers were struggling with ways to do this, particularly in math and science, expressing concerns that the mandated curriculum standards did not allow such spaces. The hydrosphere unit offered a great opportunity to see how/if curriculum from a social justice perspective (i.e., one where all students can see themselves represented) was enacted in a local context. In our initial brainstorming for the unit, Judy began Internet searches for scientists from the students' countries and posted the links on our shared Google docs. These could be used as examples as students explored questions such as, "Who studies the water cycle?" This led Tina to search for and find video clips that connected the target content/concepts with particular places (e.g., a clip on the drought in Puerto Rico, the water shortage in Malaysia).

Not surprisingly, the students responded enthusiastically to seeing and hearing the different countries and cultures as legitimate resources in science curriculum. When asked to add to the list of countries to include in a particular search, they would shout out their countries as well as others.

Patrick[3]	Dominican! Do you have Dominican?
Tina:	You didn't tell me Dominican. We'll add that.
Andrés:	Honduras! (3/8/16)

A few minutes later, William asked "What about Russia?" (ibid.)

Tina was initially skeptical of this explicit integration of students' cultures into the curriculum thinking that Judy meant students would be forced to identify only with their country or region of the world. In fact, Tina had interpreted the idea as forcing students to investigate scientists from their home countries/cultures. This was brought to light when William asked if he had to look for a scientist from Honduras. In our shared journal, she wrote,

> *"I think I cannot impose a cultural identity or what that means to them, so I allowed the students to become familiar with any scientists that they wanted to.... I do not want to artificially impose the cultural connection here."* (Google docs, 2/6/16)

Including students' cultures and countries did not mean restricting them or discouraging them from learning about other places but this is how it was interpreted by Tina and William. We include this miscommunication between us because it raised critical questions regarding

[3] Student names are pseudonyms.

intentions and implementation. This inclusive but expansive view is akin to what Joves et al. (2015) argue for in adding funds of identity to a funds of knowledge approach: affirm what students bring but allow and encourage them to build something new.

Allowing and encouraging the use of students' L1 was also an important part of the learning environment. Tina allowed and encouraged the Spanish speakers to translate for each other, clarify concepts, and even compare conditions in their different countries. Google translate was an invaluable resource for Noor and Aliya, the two Arabic speakers. A couple of the Spanish speakers, Alejandro and Vilmarie in particular, were very good at using cognates in Spanish to understand the science vocabulary in English. Tina is a monolingual English speaker but has been picking up Spanish from her students—enough so that she can sometimes answer a question in English that she hears a student ask in Spanish.

Portraits of Next Generation Scientists

To illustrate some of the rich learning that occurred within the unit we present portraits of four different students and some examples of their learning and indications of their development as next generation scientists.

Aliya: Using L1 as a resource in science learning. Aliya, originally from Sudan, arrived from Egypt in October. She exhibited a strong desire to develop content knowledge in English and her family wants her to succeed in school. She brought her science journal home every day to translate her quick write and notes from the day as well as think about her learning. Her notes and workbook pages were filled with both Arabic and English.

Early on in the unit Aliya came to school with a model of the water cycle she had created and gave it to Tina but provided no explanation. It was a piece of notebook paper glued to piece of recycled cardboard from a dry goods box. On the paper was the sun, a mountain, a tree and water cut from colored paper, two cotton balls colored blue to represent clouds, some raindrops drawn as if they were falling from the clouds and three orange arrows, pointing from water to cloud, cloud to cloud, and cloud back to water. Each part was labeled with the corresponding water cycle vocabulary: evaporation, condensation, and precipitation. The illustrated model generated good questions for us because we interpreted it very differently. Did Aliya understand the concept of the water cycle and each of the stages or was she just copying from something she saw? Tina thought Judy was not acknowledging the rich learning present and thereby not focusing enough on the positive while Judy was concerned that Tina was exaggerating the learning and worried about low expectations for students like Aliya. One thing was certain: Aliya was making a conscious effort to express her science learning to her teacher. Perhaps sharing our differing interpretations of the model made each of us attend to Aliya's subsequent work a bit more closely.

As the unit progressed, so did Aliya's development in content knowledge and English. The daily routine she created for herself was using Google translate to get the gist of both the daily question as well as questions asked orally as part of the instruction in the classroom. Aliya also worked collaboratively with her Arabic-speaking classmate, Noor. The two girls speak different dialects of Arabic but can understand each other and help each other understand the content through discussions in L1 and through their interpretations of Google's Arabic translations. Aliya is eager to communicate in English and would often volunteer to read the question of the day and/or her daily journal entry aloud. This is important to note, because teachers who are wary of allowing students' L1 in the classroom often say it is because it impedes English language development or it represents off-task behavior. Neither was true for Aliya or Noor. Adolescent ELLs have double the work (Short and Fitzsimmons 2007) of their L1 English-speaking peers in that they must develop both academic content and concepts while also developing an additional language. But when learners develop conceptual knowledge in their L1 it is easier for them to transfer or express it in their L2.

Eight weeks into the unit and only five months after Aliya came to FMS she was able to independently compose a letter to her dialogue journal partner about her learning in science. Here is one of Aliya's dialogue journal entries:

Dear Angel,

My name is Aliya but all my friends tell me Aliah. I love to draw too. I am from Sudan, I am speak Arabic and little English. This week we talked about the water crisis. I learned about dirty water and India, the problem is filters don't clean water, Of all the dirty water that is filtered, half of the water used for farming and is wasted.

Your science partner

Aliya

Aliya valued this communication time with her L1 English dialogue journal partner and was able to communicate some of the facts gathered through research about the world water crisis. She was able to write independently about her science learning with some degree of academic language. Aliya's letter shows she is addressing the guiding questions for the four-week mini-unit. She has identified a water problem and a problem with a solution: lack of clean drinking water and faulty filtration system. The notes from her science journal show that she was considering information from several countries (e.g., Bangladesh) but chose to share India in her writing journal (See Fig. 7.1). She drew several versions of her prototype for a water filter. The one in her notebook had Arabic labeling (See Fig. 7.2) and the lab sheet in her folder had her English notes. Allowing Aliya to use her L1 and L1

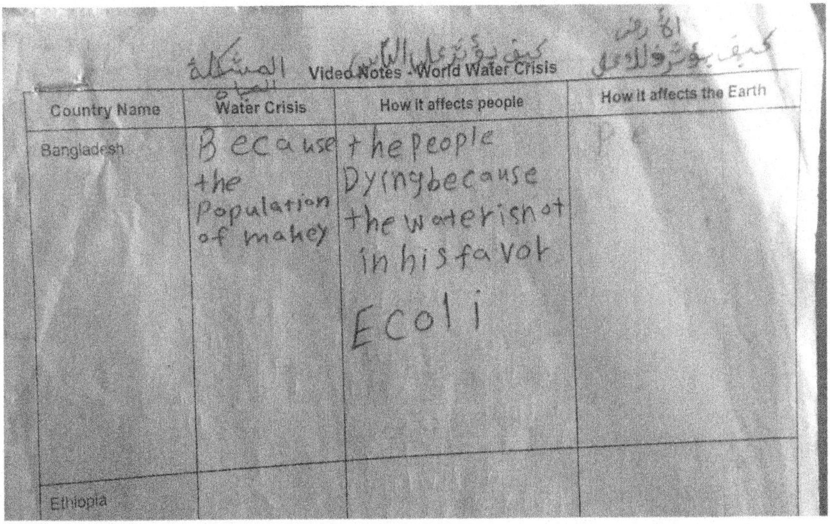

Fig. 7.1 Aliya's notes

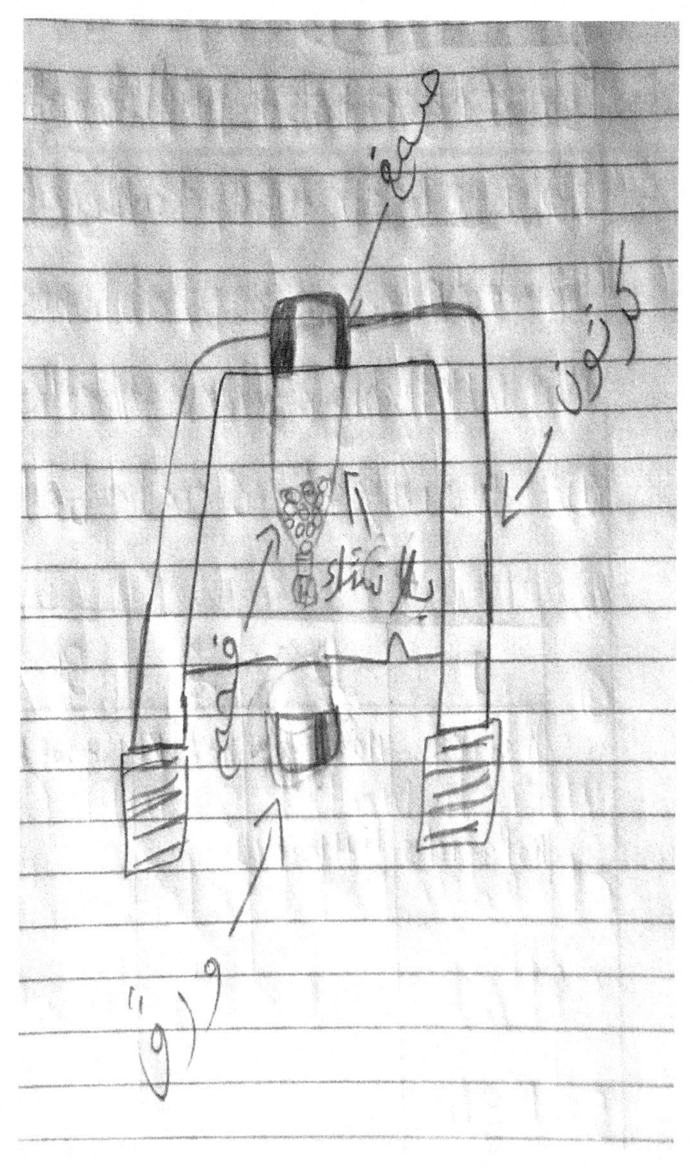

Fig. 7.2 Aliya's water filter

resources combined with the authentic communication of the journaling helped her but equally important is that they provided evidence of her learning that might have otherwise been hidden.

Alejandro: Thinking like a scientist. Alejandro arrived in September from Puerto Rico where he said he studied English in school. He was at the entering level according to the WIDA placement assessment but was progressing very quickly. By January, Tina thought he was at the developing level (3) and is confident he will be mainstreamed for grade 8. Alejandro always recognized and used the cognates that are so frequent in Spanish-English science terminology: *agricultura*—agriculture; *potable*—potable; *energía*—energy. Judy was sure this strategy allowed him to engage in science conversations much more quickly than some of his Spanish-speaking classmates who sometimes did not recognize the words until Alejandro "translated" them (i.e., said them aloud). Alejandro came to class eager to practice both the language and doing of science. He brought so much energy to discussions, questions, and explanations surrounding the potable water crisis in various countries around the world. And, he engaged in these discussions both in and out of science class. During a video describing the effects of a drought in Puerto Rico, Alejandro was quick to ask Tina if he could illustrate a water storage unit on the whiteboard. He drew a model to explain to the class the type of unit necessary to store the amount of water needed to meet basic water needs during a drought. Towards the end of the unit, when Alejandro was thinking about his water filter project, he would often seek out Tina during between class periods and talk about his design. When looking at Alejandro's water filter tests, he had made various diagrams of potential water filter designs and re-designs based on the data he collected. Each diagram of his water filter had different materials as well as different placement of materials (See Fig. 7.3). This evidence clearly illustrates that during the hands on component of this unit he was applying both previous knowledge from the water cycle lesson as well as new knowledge from the data collected during trials being run in relation to the engineering and design process. Alejandro was meeting two of the NGSS engineering design standards listed in the unit. And, the hands on/minds on (Wiggins and McTighe 2006)

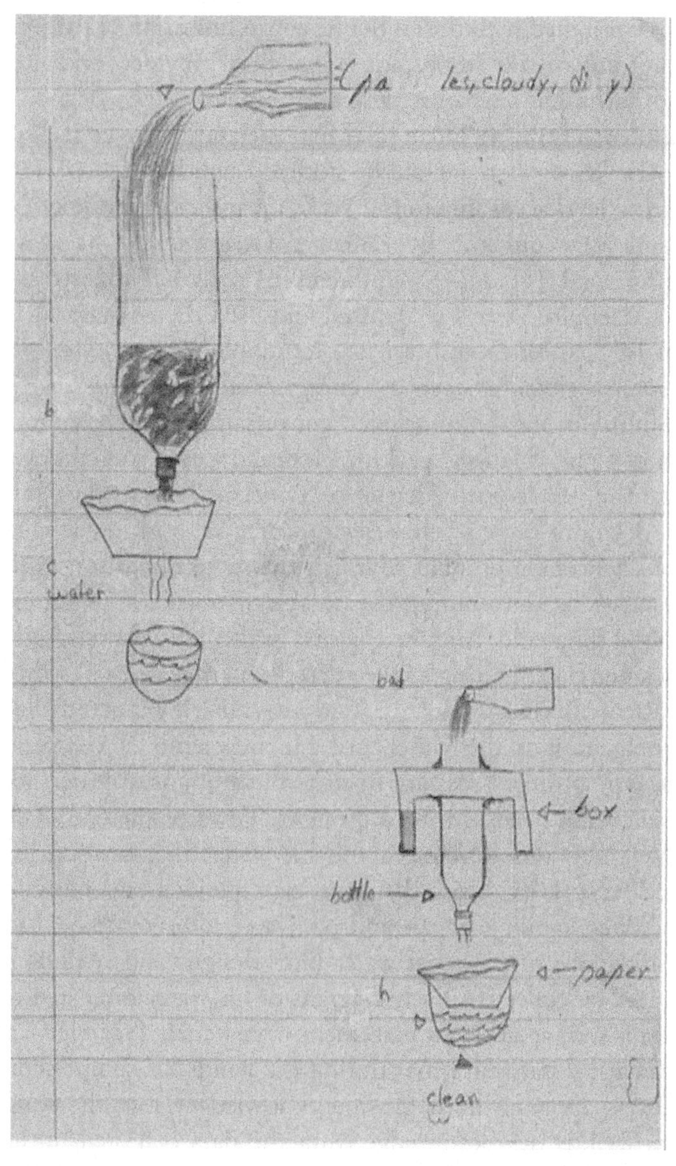

Fig. 7.3 Alejandro's filter

learning activities were supporting his acquisition of academic language, a deep understanding of content, and the ability to think like a scientist.

Anwar: "Miss, I don't read, I don't write." Anwar joined the class in January. Originally from Somalia his home languages are Somali and Maay and he is not print literate in either language. Understandably, Anwar viewed writing about science as frustrating and lacking a purpose. However, Tina could immediately see how smart he was and thought if she could build his confidence in daily oral exchanges, they could find a way for him to successfully and meaningfully participate in science class and in the writing partner activity. At first, Tina thought the model she provided for the journal would help but Anwar's partner had difficulty understanding the first letter.

MarCH 72016
Dear Jackie
My name is ANWAR I'am in mr's proulx's grade science class
We are stading about the water cycle i cmn from somala

Tina could see this was a frustrating activity for Anwar. Using pictures, gestures, and adjusted speech, Tina began to have daily oral conversations with Anwar outside of science class. Finding 5–10 minutes a day between class periods, or before/after school, Tina focused on science learning. Then, she would strategically draw on these conversations in the subsequent class so Anwar could participate more with his classmates. One of the video clips used in class focused on a clean water project involving the Maasai in Kenya. Anwar was extremely excited about the clip and was able to convey his experiences collecting water on a daily basis and what it was like living in a community with limited water resources. Anwar was developing more and more confidence in Tina's classroom but was still anxious about the writing partner assignment, sharing "Miss, I don't read, I don't write." Tina then turned their daily oral exchanges into a scribing exercise. As he spoke, Tina wrote his words and he could see that his thoughts were quickly filling up a page. He would take the page and copy it in a notebook. Anwar started to get excited about the writing exchange and checked back with Tina several times to ask for another round of oral rehearsal. Anwar was ready and excited to send the next letter, this time, complete with all the thoughts and knowledge he had shared in conversations with Tina.

Dear Jackie
I like the music i listen on computer i liked meeting you.
People need water To live. people no have water they die. Water
Is life for tor all people. in Africa people die for no have clean water
have river and Play football because there is no water and no water comes
Month comes and no water comes from sky Ten months
and no water comes no clean Water people die
then after ten months water comes but
Water cost money in Africa clean water cost money
the Government to Africa puts a group place to get water
people take a jug and fill it with water, but still Not clean
Anwar

Anwar's second letter shows he was gaining confidence in his content knowledge and his developing print literacy skills. He entered the class with concerns that he would not be able to keep up with the material. Most of his learning had been constructed through oral/auditory, visual, and kinesthetic modes, unlike the reading/writing dominated classrooms in his new culture. Anwar did not have the same L1 and technology resources or fluency of Aliya and Alejandro. Because his developing print literacy skills did not accurately reflect his content knowledge, Anwar could lose out on opportunities to engage in grade level content curriculum. Reading Anwar's second letter we see indications and evidence of complex understanding of the water crisis. He knows that water is vital to human survival but that it must be clean water. He identifies the interconnected causes of people's lack of access to clean water: weather, contamination, poverty, and inadequate government response. Anwar represents a critical challenge for teachers and schools. How to see students like Anwar as next generation scientists? Academic literacy is absolutely necessary for his ongoing education but too often ELLs are barred access from cognitively engaging content classes until they have an intermediate or advanced level of English. Yet, the hands on/minds on activities integral to inquiry-based science curriculum afford the rich, contextualized learning opportunities that foster academic language and literacy.

William: Making real world connections. William arrived from Honduras in September and entered the class with an enthusiasm for understanding the world around him. He had a successful schooling experience and was eager to share his previous knowledge and to continue learning. He was always asking questions. Several times he asked Tina, "is there a word for when a lake goes dry?" An avid follower of current events, William often offered connections between content and the news. For example, in response to a video clip highlighting potable water problems in Bangladesh, he shouted out, "Flint Michigan doesn't have clean water" (3/1/16). He appreciated examples and people from Honduras being included in science class but he was a global thinker. It was William who asked if he had to investigate a scientist from Honduras. Could he pick one from China instead? It was William who asked about Russia when Tina asked the class which countries to include when investigating water crises around the world (3/8/16).

William could express his passion and growing science knowledge orally but he was struggling with his written academic English. The partner journal was particularly beneficial for him, because it challenged him to convey meaning in a context-reduced mode (Cummins 2001; Gibbons 2015). In class, he could use gestures, drawings, and Spanish and English to convey his meaning. Sometimes when trying to explain his understanding to Tina, he and Alejandro, and sometimes Vilmarie, would confer in Spanish and then in English to Tina. He couldn't use all those resources in his partner journal. He had to depend on his ability to communicate in written English.

Learning with and from Our Students Is Inspiring but There's More Work to Do

Aliya, Alejandro, Anwar, and William all showed the potential and progress towards becoming next generation scientists as did the majority of their classmates. Their writing, speaking, and models demonstrated evidence of meeting challenging content standards while also developing English language proficiency. Moreover, they exhibited passion and enthusiasm for science. However, we are concerned that as noted by others, current formalized assessments do not adequately capture this learning and that prevailing monolingual English-only literacy perspectives can foreclose important learning opportunities for these students (Mislevy and Durán 2014).

We began our inquiry interested in knowing how integrating writing and students' cultural and linguistic identities into curriculum could help ELLs achieve the challenging standards of NGSS and CCSS. Observing the rich learning and engagement that occurred with this group of 14 students has been enlightening and encouraging. But, it has also raised several important questions and challenges.

Although the quick writes and partner journals were effective in supporting the students' learning, their successful implementation was challenging. As the class doubled in size in four months, so did the time it took for sharing responses to the daily question. We started noticing that by the time Tina was helping Anwar at one end of their large classroom table, Luis and Franklin had already finished writing and, unclear as to what to do next, would start talking about other things and/or appear disengaged. Tina had to find ways to change the checking in on students' responses from all teacher-to-student interaction to more student-to-student interactions and to establish more options for students who finished responding to the quick write before classmates. Also, depending on students' level of English and their L1 background, several needed more help with their partner journals. Tina needed to find different times of the day (e.g., between classes, at the end of the day, etc.) to work with particular students, and this can put an unrealistic time burden on teachers. Moving forward, we are developing several instructional strategies that will help students successfully run their own small groups within different parts of a lesson.

A topic for another chapter is a deeper analysis of the translanguaging practices (Garcia and Wei, 2013) that students were using to construct meaning and how teachers can better recognize and leverage those practices in the science classroom. For example, although Tina freely allowed students to use their repertoires of languages when they were working independently, she didn't highlight the different metalinguistic strategies they were using or highlight the value of the presence of multiple languages in large group discussions. This was in contrast to the ways that content connected to places was shared and valued across the class (e.g., everyone would talk about Bangladesh, China, Puerto Rico). Yet, it was fascinating to see and hear how the students were using all their different linguistic repertoires and resources to create meaning. Sometimes Aliyah and Anwar were using dialects of Arabic and/or from East Africa to communicate ideas; Noor and Vilmarie were seen using Google translate across three languages: Spanish, Arabic, and English. Alejandro consistently recognized and used cognates in Spanish, but highlighting this as a metalinguistic strategy and resource was not part of the class. Again, this is a topic that warrants further inquiry.

Final Takeaways

We are convinced that the combination of integrating strategic and structured writing activities that were ongoing, varied, and purposeful and valuing students' rich linguistic and cultural identities as learning resources and not barriers were integral to the students' achieving the challenging science standards. The young scientists in Tina's classroom are wonderful reminders that rigorous curriculum can also be humanizing (Freire 1988), cognitively engaging, and affirming (Cummins 2001), as well as socially just and joyful (Nieto 2010).

References

Cochran-Smith, M. (2004). *Walking the road: Race, diversity, and social justice in teacher education.* New York: Teachers College Press.

Comber, B. (2016). *Literacy, place, and pedagogies of possibility.* New York: Routledge.

Compton-Lilly, C., & Stewart, K. (2013). "Common" and "core" and the diversity of students' lives and experiences. In P. Shannon (Ed). *Closer readings of the common core* (pp. 63–70). Portsmouth, NH: Heinemann.

Cummins, J. (2001). *Negotiating identities: Education for empowerment in a diverse society* (2nd ed.). Los Angeles, CA: California Association for Bilingual Education.

de Jong, E., & Harper, C. (2005). Preparing mainstream teachers for English language learners. *Teacher Education Quarterly, 32*(2), 101–124.

Echevarría, J., Vogt, M., & Short, D. (2014). *Making content comprehensible for secondary English learners: The SIOP Model* (2nd ed.). Upper Saddle River, NJ: Pearson.

Freire, P. (1970//1988). *Pedagogy of the oppressed.* New York: Continuum.

García, O., & Wei, L. (2013). *Translanguaging: Language, bilingualism and education.* New York: Palgrave.

Gee, J. (2005). Language in the science classroom: Academic social languages as the heart of school-based literacy. In R. Yerrick & R. Wolff-Michael (Eds). *Establishing scientific discourse communities: Multiple voices of teaching and learning research* (pp. 19–37). Mahwah, NJ: Lawrence Erlbaum.

Gibbons, P. (2015). *Scaffolding language, scaffolding learning: Teaching second language learners in the mainstream classroom.* Portsmouth, NH: Heinemann.

Joves, P., Siques, C., & Esteban-Guitart, M. (2015). The incorporation of funds of knowledge and funds of identity of students and their families into educational practice: A case study from Catalonia, Spain. *Teaching and Teacher Education, 49*(1), 68–77.

Kumashiro, K. (2009). *Against common sense: Teaching and learning toward social justice* (Revised ed.). New York: Routledge.

Mislevy, R., & Durán, R. (2014). A sociocognitive perspective on assessing EL students in the age of the common core and next generation science standards. *TESOL Quarterly, 48*(3), 560–585.

Moll, L. C., Amanti, C., Neff, D., & González, N. (1992). Funds of knowledge for teaching: a qualitative approach to connect households and classrooms. *Theory into Practice, 31*(2), 132–141.

National Research Council. (2012). *A framework for K-12 science education: Practices, crosscutting concepts, and core Ideas*, Committee on a Conceptual Framework for New K-12 Science Education Standards. Board on Science Education, Division of Behavioral and Social Sciences and Education. Washington: DC The National Academies Press.

National Writing Project. (n.d.). *About us.* Retrieved from http://www.nwp.org/cs/public/print/doc/about.csp

Next Generation Science Standards. (2013). Appendix M. Retrieved from http://www.nextgenscience.org/sites/default/files/resource/files/Appendix%20M%20Connections%20to%20the%20CCSS%20for%20Literacy_061213.pdf

Nieto, S. (2010). *Finding joy in teaching students of diverse backgrounds: Culturally responsive and socially just classrooms in U.S. classrooms.* Portsmouth, NH: Heinemann.

Palincsar, A., & Schleppegrell, M. (2014). Focusing on language and meaning while learning with text. *TESOL Quarterly, 48*(3), 616–623.

Peregoy, S., & Boyle, O. (2013). *Reading, writing, and learning in ESL: A resource book for K-12 teachers* (6th ed.). Boston: Longman.

Short, D., & Fitzsimmons, S. (2007). *Double the work: Challenges and solutions to acquiring language and academic literacy for adolescent English language learners—A report to carnegie corporation of New York.* Washington, DC: Alliance for Excellent Education.

Suárez-Orozco, C., Suárez-Orozco, M., & Todorova, I. (2008). *Learning in a new land: Immigrant students in American society.* Cambridge, MA: Harvard University Press.

Wiggins, G., & McTighe, J. (2006). *Understanding by design* (2nd Ed.). Upper Saddle River: NJ Pearson.

Judy Sharkey is Associate Professor in Education at the University of New Hampshire, Durham NH. Her research focuses on teacher learning and development in plurilingual/transmigrant communities and is informed by critical sociocultural learning theories and social justice teacher education (SJTE). Since 2009 she has been involved in a collaborative international research project investigating the role of community-based pedagogies with teachers and colleagues in Bogotá, Colombia. She has directed two multiyear federally funded professional development projects designed to enhance the academic experiences and opportunities of NH's bilingual/multilingual students. A former Peace Corps Volunteer and Fulbright scholar, she has edited two books highlighting the role of teacher knowledge in curriculum development, and her research has appeared in *Journal of Teacher Education, TESOL Quarterly,* and *Curriculum Inquiry* among others.

Tina Proulx is a public school science/ELL teacher in a program designed for newly arrived immigrants and refugees in Manchester, New Hampshire. She has a B.A. in Education, an M.Ed. in Teacher Leadership and a CAGS in

Educational Leadership/Curriculum and Assessment. She is also adjunct faculty at the University of New Hampshire where she teaches and supervises in the ESOL certification program. Tina has worked to form partnerships among local institutions of higher education and her school district to create and manage an after school program for students and their families. She also works closely with family bilingual liaisons to promote community awareness.

8

Helping English Language Learners Access the Language and Content of Science Through the Integration of Culturally and Linguistically Valid Assessment Practices

Sultan Turkan and Alexis A. Lopez

In this chapter, we discuss culturally and linguistically valid classroom assessment practices[1] that can help English language learners (ELLs) learn both science and the language of science. We target science teachers and teacher educators as our audience. To describe these assessment practices, we draw on the interpretive validity argument (Kane 2006) and cultural validity perspective (Basterra et al. 2011). We propose that the existing culturally and linguistically valid assessment practices could be put into formative use in science classrooms to help ELLs access the language and content of science as guided by the Next Generation Science Standards (NGSS) (NGSS Lead States 2013). Access to language and content of science also means maintaining the cognitive demands of the science content without simplification. Long

[1] By assessment practices, we refer to practices of using methods of assessment in classrooms for formative purposes.

S. Turkan (✉) · A.A. Lopez
Educational Testing Service, Princeton, USA
e-mail: sturkan@ets.org; alopez@ets.org

© The Author(s) 2017

L.C. de Oliveira, K. Campbell Wilcox (eds.), *Teaching Science to English Language Learners*, DOI 10.1007/978-3-319-53594-4_8

before NGSS, language-intensive practices in science had been emphasized (Lemke 1990). As NGSS reemphasizes language-intensive practices, we take the language functions embedded in the standards to illustrate how culturally and linguistically valid assessment practices could be brought into the classroom to support the teaching and learning of science-specific language and content.

NGSS has been adopted by 18 states and the District of Columbia. It has emphasized acquisition of literacy skills along with inquiry skills in science and deemphasized the learning of detailed facts as well as "loosely defined inquiry" (Lee et al. 2013, p. 223). Science is now depicted in terms of three dimensions: science and engineering practices, crosscutting concepts, and disciplinary core ideas. An issue that crosscuts and connects these three dimensions is *language*. The prominence of language-intensive practices in the standards definitely poses challenges and opportunities for ELLs as well as teachers charged to align their instruction with NGSS.

Both the challenges and opportunities arise from the following language-intensive practices that call for using science-specific discourse: asking questions; engaging in argument from evidence; obtaining, evaluating, and communicating information; constructing explanations; and developing and using models. In order to engage in these practices, all students, especially ELLs, need specialized support from science teachers (Lee et al. 2013), as indicated by the persistent achievement gap between ELLs and non-ELLs in mathematics and science (Aud et al. 2011). It is reasoned that ELLs have a harder time engaging in science inquiry meaningfully, because they are acquiring English language literacy and content understanding at the same time (Fradd and Lee 1999; Kelly-Jackson and Delacruz 2014).

To help ELLs engage with the scientific content and language, pre-service science teachers need tools to provide effective instruction of science content and language. We argue that assessment can help to provide opportunities for all learners to learn language and content, especially if and when it is utilized in culturally and linguistically valid ways. To develop and improve effective science instruction as well as equitable assessments, teachers should address language as an integral part of teaching science (Lyon 2013a, 2013b). Classroom assessment practices could especially help students to practice with language and use it to demonstrate content understanding. With that, this chapter is motivated by the premise that classroom assessment practices have the potential to serve as a tool to facilitate effective science teaching for ELLs.

In this chapter, we first discuss the background and research on linguistically and culturally valid classroom assessment practices framing it through the lens of equitable assessments. In the second part of the chapter, we propose an approach to developing linguistically and culturally valid assessment practices, in consideration of the research and theoretical lens reviewed in the first half.

Why Should Science Teachers of ELLs Care about Culturally and Linguistically Valid Classroom Assessment Practices?

One of the tools that classroom teachers can use to make a substantial impact on science learning is assessment. Assessment plays a critical role in secondary science classrooms as it is typically used to determine what students know, monitor student progress, and inform instruction (Lyon 2013b). NGSS, along with the Common Core State Standards (CCSS) (Common Core State Standards Initiative 2010), provides a unique opportunity not only to shape instruction, but also to enhance classroom assessments as formative tools to support ELLs' learning of science content and language (Hakuta 2014). When already faced with the mandate to incorporate language-intensive instructional practices, teachers would greatly benefit from making classroom assessment practices more linguistically and culturally valid to engage ELLs in the learning process.

In terms of making assessment practices valid, we are informed by the theories of validity prevalent in the field of assessment and measurement. Kane (1992, 2001) argues that the claims made about student performance should be based on valid test score interpretations. In order for tests to yield valid score interpretations, one view is that the knowledge and skills elicited from students should not include variance irrelevant to the construct being assessed, that is, construct irrelevant variance (Messick 1989). Some have taken this premise to argue that in order to draw valid score interpretations from the performance of ELL students, language of the content being assessed on content assessments should be modified or simplified (Abedi et al. 2004). Some took the view to argue that ELLs should be given the fair opportunity to demonstrate their content knowledge in the context of assessing their disciplinary knowledge in a field like science (Lyon 2013a, 2013b). This view has been framed through the equitable assessments lens which essentially aims to promote integrating various ways of eliciting student understanding (Siegel 2007; Solano-Flores and Nelson-Barber, 2001).

With that, science assessments should provide equal opportunities for all students, including ELLs, to demonstrate what they have learned in the science classroom regardless of their cultural and linguistic background and their English language proficiency (Buxton et al. 2013; Lyon 2013a, 2013b; Noble et al. 2014; Siegel 2014). ELLs bring diverse cultures, epistemologies, and lived experiences to the classroom that shape how they view, learn, and communicate their understanding of science (Solano-Flores and Nelson-Barber 2001; Turkan and Liu 2012). In order to enhance the science learning process, teachers should embrace this heterogeneity that ELL students bring to the science classroom. Therefore, it is critical for science teachers to develop assessment practices that allow ELLs to demonstrate their understanding of science knowledge in equitable ways.

This idea has been captured and discussed within the purview of equitable science assessments which can be defined as assessments that give all students an opportunity to demonstrate understanding of science knowledge (Pullin and Haertel 2008). We employ this lens to synthesize the research on linguistically and culturally valid assessment practices in science classrooms.

Equitable Assessments for ELLs

Equitable science assessments tap into ELL students' experiences, culture, and language to give them opportunities to demonstrate what they have learned in the science classroom and allow students to demonstrate understanding in multiple ways (Lyon 2013b). Equitable assessments have the potential to allow science teachers to make more informed decisions about what ELL students are learning and thus determine what type of support ELL students need in order to inform and plan future instruction.

The notion of equitable assessments evolved over the years in relation to the role language played in the interpretation and demonstration of content knowledge as elicited on content assessments. Earlier, it was suggested that unnecessary complex linguistic features of the assessment should be modified to alleviate the interference that such language might pose to ELLs' demonstration of content knowledge. This view has long dominated the research on fair and valid assessment accommodations for ELLs (Abedi and Lord 2001; Bailey 2007; Young 2009). In this line of research, it was found that language demands of the academic content are challenging for ELLs and impact their performance on content assessments (Wolf and Leon 2009; Martiniello 2008). Therefore, various linguistic and non-linguistic accommodations have been employed in the field (Pennock-Roman and Rivera, 2011). The most recent shift in making content assessment equitable and linguistically responsive has been towards positioning language as a tool and resource. In that, language serves to "*value* and *scaffold* students' use of language in assessment by modeling language use, providing visual support, and making the assessment content relevant to student culture and the local and the physical environment (Siegel 2007; Stoddart et al. 2010)" (Lyon 2013b, p. 3). With this formative potential, equitable assessments could incorporate various linguistic and non-linguistic scaffolds that "might help the student comprehend the question, think about the topic, or respond to the prompt" (Siegel 2007, p. 867). In terms of non-linguistic scaffolds, the claim is that equitable classroom science assessment practices should be culturally valid to students' lived experiences at home and in the community (Lee et al. 2011) in that they should account for ELL students' cultural norms and ways of knowing in culturally sensitive ways (Lee 2003).

Lyon's (2013b) study of how three science teachers evolved in their expertise and understanding in equitable science assessments shows that it may be possible to bring equitable assessment practices to life in classrooms. For example, one of the participating teachers started off by focusing on the content of the assessment only and overlooking the role of language. His understanding then changed towards scaffolding students' writing of scientific explanations by providing charts for students to complete in response to specific guiding questions. This scaffold involved engaging students in sustained dialog around the language function of explaining. Another teacher changed her understanding of equitable assessments by providing various forms of assessment, such as describing and drawing, explaining verbally and in writing. The third teacher utilized language frames as well as multi-step directions for guiding students through a lab. Overall, all three teachers provided learning opportunities for ELLs by allowing them to engage in *assessment conversations*, such as written predictions, group discussion around prompts, and scientific explanations whereby ELLs had to use some form of science discourse. The case studies pointed to a tension that teachers experienced about whether to reduce language demands of the assessment or scaffold the language demands by allowing students to demonstrate their knowledge in multiple ways. Another tension was in assessing language use in addition to conceptual understanding. Overall, Lyon (2013b) concluded that it is possible for science teachers to learn how to assess language while assessing science knowledge.

Inspired by these applications and informed by the theoretical underpinnings of equitable assessments, we next review culturally and linguistically valid assessment practices in science and we recommend practices that science teachers can implement to help ELLs learn the language and content of science. First, we describe the characteristics of linguistically valid assessment practices in general and particularly, in science. Then, we explain what it means to integrate culturally valid assessment practices into instruction.

Linguistically Valid Assessment Practices

All content assessments depend on language in some way (Trumbull and Solano-Flores 2011), but integrating science and language poses a challenge and an opportunity. Despite the challenge, science assessments have the power to promote science and language learning for ELLs, especially when ELLs are supported to use language as a tool to perform complex multi-dimensional tasks (Lee et al. 2013). Here, we highlight three linguistically valid science assessment practices from the literature to draw the approach proposed in this chapter.

First, the scientific discourse, which refers to ways of knowing, constructing, and communicating content in science, should be incorporated in the assessment (Lyon 2013a). The use of scientific discourse in science assessments has the potential to promote ELL students' scientific literacy as it encourages them to use complex thinking and gives them the opportunity to use the language of science to demonstrate science knowledge (Lyon et al. 2012). Science assessments that emulate authentic scientific discourse tend to elicit more student thinking rather than simply factual knowledge (Lyon 2013b). Science assessments that allow students to explain their responses and to clarify their thinking provide teachers with rich information about how their students interpret science and how they use language to demonstrate understanding of science concepts (Buxton et al. 2013; Lyon 2013a).

This means that the use of scientific discourse should be viewed as part of the construct in science assessments (Avenia-Tapper and Llosa 2015; Buxton et al. 2013). To achieve this, two components of content-specific discourse (Schleppegrell 2001) are worth drawing on for our purposes here. One is the linguistic functions that are specific to scientific discourse within a particular science content area. The second component is the set of linguistic features associated with these functions. We define linguistic functions as functions that serve to express meaning through using language. Examples include classifying and comparing/contrasting; describing, explaining, and elaborating; giving/following directions; making generalizations; predicting; summarizing; sequencing; expressing needs, likes, and feelings; expressing cause/effect; drawing conclusions; clarifying; proposition/support; and the like (Dutro and Kinsella 2010, p. 171). These linguistic functions are realized through the use of linguistic choices and features that differentiate the language of a particular content area from the language of another content area. Some of the science-specific linguistic functions that are covered across the NGSS scientific and engineering practices are as follows: explaining, describing, providing evidence, formulating a problem, building arguments, summarizing, sequencing, classifying, comparing and contrasting, making generalizations, predicting, and so on. The linguistic features and choices depend on the particular functions students are asked to perform.

In relation to linguistic features, science makes unique use of lexicon, semantics, and syntax (Fang 2006). At the lexical level, science uses technical vocabulary. At the syntactical level, the use of features such as nominalization, the use of passive voice, interruption constructions, and ellipses are observed (Fang 2006). To briefly describe each feature, nominalization helps to economically condense complex predicates into abstract noun phrases such as "the tumbling and splashing of swiftly flowing water" exemplified in Fang (2006, p. 500). The use of passive voice is common in the language of school science, because subjectivity is avoided in favor of the perceived objectivity of the passive voice. Interruption constructions refer to grammatical structures such as adverbial clauses serving to modify the expected clause pattern. Fang (2006) identifies the use of "many too small to be seen" as an *interruption construction* in the following sentence: "Hundreds of pores, many too small to be seen without a hand lens, dot a sponge's body" (p. 504). Ellipsis, or the omission of words, phrases, or clauses, as a mainstream feature of written English is commonly identified in science texts to eliminate redundancy. The phrase "able to open the tough pods" at the end of the following sentence is an example for ellipsis: "Finches with larger and stronger beaks were better able to open the tough pods than were finches with smaller, weaker beaks" (Fang 2006, p. 497).

Second, instead of simplifying the language demand of the assessment tasks, valid science assessments can use different types of scaffolds to support ELL students in understanding the tasks and also in demonstrating their science knowledge and what they have learned in the science classroom (Siegel 2014). Scaffolds serve as a resource to support ELL students in understanding and using the language of science in the assessment. Scaffolds could take various shapes and forms, depending on the language function and features that the teacher aims for. One of the important functions that scaffolds should serve is to make the assessment task reflect the characteristics of the students (Mislevy and Durán 2014). In other words, a task will be matched to students who will be familiar with the content, terms, and language of the task. Otherwise, the task will not be appropriate for the students.

Third, ELL students should be allowed to use both everyday language and the academic language of science and to demonstrate their science knowledge in multiple ways (Buxton et al. 2013). Using multiple measures to assess ELL students' science knowledge will ensure that ELL students can demonstrate their science knowledge in various methods, including speaking, writing, and drawing (Kopriva et al. 2013; Noble et al. 2014). It is important to provide various opportunities for the students to respond and demonstrate their knowledge about science rather than the multiple-choice format (Noble et al. 2014). Noble et al. (2014) found that ELLs made alternative interpretations of a science item in the multiple-choice format and ended up responding to a different question. In doing so, however, they demonstrated that they have the targeted knowledge in science.

Culturally Valid Assessment Practices

Teachers should cultivate cultural competence to appreciate how individual students respond to instruction and assessment differently, because students' responses are influenced by their cultures (Basterra 2011). For example, American Indian/Alaska Native students might be more comfortable with open-ended formats rather than single answer formats, such as multiple-choice and true/false (Basterra 2011). Similarly, these students might prefer collective questioning approaches over the individual classroom questioning. Hence, students from diverse backgrounds might interact with assessments in diverse ways.

When it comes to enacting culturally sensitive assessments, teachers should essentially allow ELL students to draw on their cultural and linguistic resources (Fusco and Barton 2001; Lee and Fradd 1998; Solano-Flores and Nelson-Barber 2001; Warren et al. 2001). This can be accomplished by activating ELL students' prior knowledge about science and by recognizing how they learned science (Warren et al. 2001). Another way to make science assessments culturally sensitive is by situating the assessment within culturally meaningful contexts to make the assessment content relevant to the students' culture and the local and physical environment (Siegel 2007). As is generally the case, it is also important to provide different types of assessments to assess the intended knowledge and skills (Basterra 2011).

Specifically, research has shown that performance-based tasks, such as open-ended questions, allow students to use and practice the scientific discourse and to clarify their thinking (Buxton et al. 2013). Conversely, research has also shown that the use of multiple-choice questions to assess science is challenging, because these types of items do not provide teachers with as much information about their students' science knowledge and about how they use scientific discourse as more open-ended tasks do (Buxton et al. 2013). Other studies have shown that some ELL students have difficulties interpreting multiple-choice questions appropriately (Noble et al. 2014).

In relation to making science relevant to ELLs' lives, Lee and Fradd (1998) describe how the gaps between students' everyday experiences on the one hand and academic concepts and classroom practices on the other might be overcome through "the process of mediating the nature of academic content with students' language and cultural experiences to make such content (e.g., science) accessible, meaningful, and relevant for diverse students . . . " (p. 12). Lee and Fradd (1998) coin the term *instructional congruence* to characterize this process. Drawing on Lee and Fradd's concept of instructional congruence, Moje et al. (2001) argue that the potential conflict between science and classroom practices and students' everyday language and literacy practices can be lessened when teachers supplement the textbook by allowing ELLs to bring knowledge of scientific concepts from their everyday lives to the classroom. Their study highlights the importance of instructional congruence and of involving linguistically and culturally diverse students in spoken and written scientific discourse by making connections to students' home cultures and everyday experiences.

Proposed Approach to Developing Culturally and Linguistically Valid Assessment Practices

Taking into account the research and theoretical lens reviewed earlier, we can describe the considerations that science teachers can use to develop culturally and linguistically valid science assessment practices in their classrooms. We propose that teachers follow three steps to implement the considerations. These steps are illustrated with examples in Table 8.1.

The NGSS lend themselves to the specific language functions that each topic suggests, and the first step for teachers is to link the NGSS disciplinary core idea and respective science or engineering practices to

Table 8.1 Three steps in developing culturally and linguistically valid cultural assessments

Proposed three steps	Breakdown of the three steps	Example
1. Link the NGSS disciplinary core idea and respective science or engineering practice to the relevant language functions and objectives explicitly or implicitly stated by the science or engineering practices	Identify the language function(s) embedded in the science or engineering practice	Providing evidence or support (science and engineering practice of engaging in argument from evidence)
2. Identify the applicable assessment method that would allow for accessing the content and demonstrating knowledge and understanding of the content	Align the language function with the selected assessment method	"Explain why" task
3. Make the task linguistically and culturally valid using various types of scaffolds	Consider the target language function, linguistic features, and characteristics of the students	Using native language as support, connecting to prior knowledge, and linguistic unpacking

the relevant language functions and objectives explicitly or implicitly stated by the science or engineering practice. As mentioned in Table 8.1, this step involves identifying the language function(s) embedded in the science or engineering practice. Let us recall that a language function basically refers to a function or purpose for communication through reading, listening, writing, or speaking. Depending on the specific science or engineering practice, the language function could be one or two of the following: explaining, describing, and elaborating; providing evidence or support; formulating a problem; making an argument; summarizing; sequencing; classifying; comparing and contrasting; making generalizations; predicting, and so on. For example, one language function that would link to the practice of engaging in argument from evidence is providing evidence or support for claims.

The second step is to identify the applicable assessment method that would allow ELLs to (a) access the content and language of the assessment and (b) demonstrate their knowledge and understanding of the scientific content and practice elicited in the assessment. The assessment method depends on what performance expectation the teacher would like to focus on as well as what scaffolding the teacher chooses to embed in the assessment. Again, identification of the language function is key here for full alignment with the assessment method. Figure 8.1 shows the various types of assessment methods, such as discrete and performance, which would allow ELLs to access the assessment and to demonstrate their knowledge. Performance strand includes two main types of classroom assessments: written and oral. Written assessments could include, but are not limited to, the following assessment tasks: (a) construct an explanation using the evidence, (b) write a lab report, and (c) answer specific questions with short sentences and the like. Oral assessments include, but are not limited to, the following tasks: (a) explain to your peer using evidence, (b) critique a model with your peer, and (c) present orally to the class on the topic of assignment. As illustrated in fig. 8.1, discrete assessment refers to a set of items or prompts that require single responses from the students. This assessment method could take various forms, such as (a) single selection (i.e., selecting from multiple options provided by the teacher), (b) single selection and explain the reason for your selection, and (c) fill in the

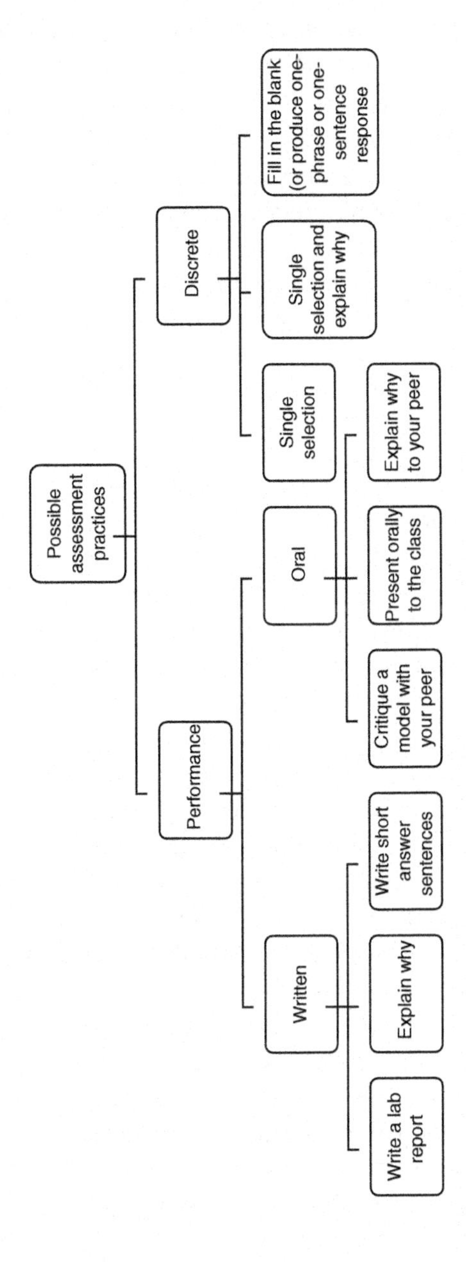

Fig. 8.1 Three steps in developing culturally and linguistically valid cultural assessments

missing information (i.e., blank). To reiterate, discrete assessments could be used in combination with the performance assessments.

Third, the task would provide scaffolds that would help ELLs access the language and content of the assessment and demonstrate their understanding of the content. In a classroom assessment environment, scaffolds could take various shapes and forms, such as the teacher reading the question aloud for the student; explaining words, phrases, and sentences; using the native language as support; connecting to prior knowledge; and drawing pictures, graphs, or tables.

We explain each of these steps next illustrating with the example scenario.

Application

Here, we present an assessment task illustrating the three main steps on a 6–8th grade science standard. The goal of the task is to assess 8th grade students' abilities to construct scientific explanations that are supported by multiple sources of evidence. In this example, the larger NGSS core idea is *Ecosystems: Interactions, Energy, and Dynamics*. Within this core idea, the class is beginning to work on different types of species' interactions including predation, competition, and symbiosis. The specific topic that is being assessed is interdependent relationships in ecosystems, and the type of interaction could be competitive, predatory, or mutually beneficial. The teacher (Ms. Billig) in this class wants to focus on developing the scientific knowledge, the scientific practices, and the discourse features related to this topic. The main science practice that is being assessed is constructing explanations supported by multiple sources of evidence to predict phenomena. Other scientific practices associated with the standard are developing models to describe phenomena and analyzing and interpreting data to provide evidence for phenomena. The crosscutting concept associated with this standard is recognizing and using patterns to identify cause and effect relationships.

In order to situate this classroom scenario within the three steps mentioned before, we need to first identify the language function embedded in the science or engineering practice that is of focus for the teacher. In this example scenario, since the performance expectation is to have students "construct an explanation that predicts patterns of interactions among and between organisms" (NGSS Lead States 2013) using data from a specific ecosystem, the two specific language functions are explaining and making claims. Depending on the language function(s) that the teacher has emphasized in the classroom, the assessment method could be selected either from the performance or discrete strand or both strands used in combination. In this classroom assessment scenario, Ms. Billig chooses to have students construct a scientific explanation in writing. As scaffolds, though, she considers various options to provide discrete and open-ended opportunities for ELLs to understand the task as well as demonstrate their understanding of content.

Description of the Assessment Task

In this task,[2] Ms. Billig asks students to construct an explanation that predicts a pattern of interaction between the white-tailed deer population and the gray wolf population in a forest reserve. One of the purposes of this assessment task is to allow students to use authentic scientific discourse or argumentation. As part of the task, the students are provided with a set of data about the population size of wolves and deer, as well as deer deaths due to predation, in the last ten years (see Table 8.2). The information is presented in both text (e.g., short reading passage) and graphical (e.g., line graph and table) form.

Identifying Language Functions and Features Associated with the Standard

Since the first step is to identify the language function embedded in the science practice, the main language function in this task is to explain a natural phenomenon. Also embedded in the science

[2] We adapted the example task here with input from two sources: 1) https://www.wolfquest.org/pdfs/Deer%20Predation%20or%20Starvation%20Lesson.pdf https://www.biologycorner.com/worksheets/predator_prey_graphing_key.html

Table 8.2 Number of gray wolf population and white-tailed deer population from 2006–2015

Year	Gray wolf population	White-tailed deer population	Deer deaths due to predation
2006	14	2,500	210–280
2007	16	2,800	240–320
2008	20	3,000	300–400
2009	26	2,850	390–520
2010	32	2,710	480–640
2011	28	2,600	420–560
2012	25	2,550	375–500
2013	22	2,600	330–440
2014	23	2,580	345–460
2015	23	2,588	345–460

practice of constructing a scientific explanation is making a claim based on patterns of interaction. In order to construct the scientific explanation, students need to write a claim and then write a reasoning statement connecting the evidence to the claim. To write the claim, students are taught to use the linguistic choice *conditionals*. The key vocabulary in the task are words that have already been presented in class (e.g., population, predator, prey, predation, increase, decrease). Students also need to be familiar with the irregular plural forms of the words *wolf* and *deer*.

Identifying the Assessment Method

Ms. Billig then chooses the assessment method that would allow students to access the content and demonstrate their understanding and knowledge of the content. Ms. Billig thinks that performance tasks are most aligned with the targeted language function and therefore decides to design a task that asks students to construct a scientific explanation in writing. Following is the assessment task Ms. Billig developed for her class.

Prompt: Read the following text about the wolf and deer populations in a forest reserve. Based on the data presented in the table below, you will need to provide an answer to the following research question: How do these populations interact within this environment? To address the question, please construct a scientific explanation that predicts the pattern of interaction between the white-tailed deer population and the wolf population in this forest reserve.

Gray Wolves and White-tailed Deer in a Forest Reserve

The Department of Natural Resources (DNR) protects and regulates the area's wildlife resources. In 2001, the DNR created a program to monitor the gray wolf and white-tailed deer populations in the area. Scientists at DNR have been gathering data about population densities of wolves and deer in order to understand how they interact with their environment. The table below shows some of the results from this program for the last ten years.

Making the Task Linguistically and Culturally Appropriate

As a third step, Ms. Billig wants to ensure that the task offers affordances for ELL students who might not have sufficient English language proficiency to complete the task. Thus, Ms. Billig aims to make the task culturally and linguistically relevant, not just to simplify the cognitive demand of the task.

One of the ways to make the task culturally relevant is to change the context of the task. In this particular example, the teacher can use local species, or species that ELL students are more familiar with. For example, instead of focusing on interactions between the white-tailed deer population and the wolf population in the forest reserve, the students can focus on the interactions between species in their local context (e.g., Florida), focus on interaction between other types of species (e.g., rabbit-foxes, grass-rabbit, shark-seal, cats-mice), or allow students to choose to focus on species that they are familiar with, including species in their home countries.

To make it linguistically appropriate, Ms. Billig would like to provide various scaffolds targeted to help the students to understand the question or task as well as allow them to demonstrate their understanding in multiple ways, such as open-ended responses in writing or orally or through illustrations. Also, to help the students understand what the task or prompt is asking them to do, she knows she has several options: simplifying the language in the directions, translating the directions to the students' home language, reading aloud the directions, or modeling the task to the students. Alternatively, she could add illustrations (two species) or pre-teach key vocabulary (prediction, interaction, etc.).

Specifically, the scaffold type depends on what type of support Ms. Billig wants to provide for her students based on the language function she aims for. As mentioned earlier, the specific language function that the teacher is aiming for depends on the content focus and targeted scientific or engineering practice. Accordingly, the scaffolding tasks would be embedded within the main task, and all students would be required to complete them. Later, we illustrate a few sample scaffolds; each scaffold can be presented in the form of subtasks.

Sample scaffold 1. Since students are being asked to draw on data and evidence to support their claim about a pattern of interaction between two organisms in a particular ecosystem, it is very important for ELL students to understand the data and evidence presented in texts, tables, or graphs. To that end, for example, Ms. Billig could ask students to identify information in the table by answering questions (e.g., *Which population was higher in 2010, wolves or deer? In what year did the deer population start decreasing?*). Or Ms. Billig could ask students to determine if certain statements are true or false (e.g., *The population of deer was higher in 2011 than in 2009, or the population of deer started decreasing in 2008.*). Ms. Billig can help students by asking them certain key questions (e.g., *Do you notice any pattern in the wolf and deer populations over 4 or 5 years of time? Can you see any differences or similarities between the wolf and deer populations over time?*). Furthermore, students must also decide if the data they have selected in their argument is relevant, sufficient, and convincing enough to support their claim. This will also help them evaluate competing claims they may think of or that peers come up with. To this end, Ms. Billig could further support students by asking probing questions such as (a) *What are you trying to determine?*, (b) *Why is that evidence you selected important?*, and (c) *Does this data seem to fit with other data we have about predator/prey interactions?*.

Sample scaffold 2. To help students understand the data presented in the table, another scaffold/subtask that could help students to connect

Table 8.3 Scaffold 2

How do the wolf and deer populations interact in this forest reserve?	
Claim (What is your claim?)	Evidence (What evidence/data in the table supports your claim?)

their claim to the evidence/data is to provide a graphic organizer on which students are asked to provide the missing information. The first step is for the students to analyze and interpret the data. Then they can select the data they think is sufficient to best support their claim (Table 8.3).

Sample scaffold 3. In doing all this, students, especially ELLs, will need help with writing a claim to answer the research question based on the evidence/data in the table. One way to facilitate this is to have students answer a series of questions and then write the claim. To help students answer the questions, the teacher could provide language frames, and students can fill-in-the blanks to complete them. The two potential claims are illustrated next with guided language frames.

Claim A

If the population of wolves increases, what happens to the population of deer?

The population of deer ____________ because the population of wolves ____________________.

Claim B

If the population of wolves decreases, what happens to the population of deer?

The population of deer ____________ because the population of wolves ____________________.

Sample scaffold 4. Since the goal of this task is to support students in writing a scientific reasoning statement to connect the evidence to their claim, one scaffold must ensure that students attend to several linguistic features. Specifically, linguistic features that identify a reasoning statement connecting evidence to a claim might include, but are not limited to, the use of subjective modals (could, may) and causal conjunctions (because, as, since, therefore). While the relationship between deer and wolves is familiar to most students, the scientific use of causal linguistic features may not be as straightforward, especially to ELLs. Going back to our example assessment task, Ms. Billig could help link a claim to the evidence by modeling language use and providing examples such as the following sentence: *If the population of wolves increases, the deer population decreases because the number of dear death due to predation would increase.*

Alternatively, Ms. Billig could encourage them to refer back to the table and find the supporting evidence to back their claim, such as in the following two example sentences:

1. *Looking at the data, from 2006 to 2010, the gray wolf population increased, deer death due to predation also increased.*
2. *If the population of wolves increases, the deer population decreases due to death by predation. Looking at the data, from 2006 to 2010, the gray wolf population increased as did the number of deer deaths from predation*

It should be reiterated here that the scaffolds we presented for illustration purposes could take multiple shapes and forms depending on the particular linguistic and non-linguistic needs of the ELLs in the particular classroom. Regardless of the particular needs, the targeted language function(s) should guide the process of devising scaffolds. For example, if the language function associated with the particular content to be covered is explaining why in response to an audio input, then the input could be presented to the students through listening.

Summary and Implications for Teacher Preparation

Science is a social process in which scientific knowledge is defined, revised, and communicated based on evidence (McNeill and Pimentel 2010). As scientific knowledge is constructed on the social plane (Lemke 1990), communication is central to constructing knowledge in science. As a consequence, science discourse is highly language intensive.

Language is an inevitable part of making science. In teaching classroom science though, science teachers might inadvertently treat science as "a static set of facts" (McNeill and Pimentel 2010 p. 205), or they may focus on the content by deferring the responsibility to the teachers of English as a second language or English language arts teachers. Attending to language gets even more challenging for mainstream teachers, especially when and if the teachers have linguistically and culturally diverse students who are challenged with the language of science. In the current era of the NGSS, as the standards emphasize the language-intensive practices, the need to help these students meet the challenges of science is even more imminent than before.

In this context, we have argued that classroom assessment could be utilized as a lever for teaching content and language. This view essentially is guided by the equitable assessment framework that positions assessment as "a social interaction driven by instructional questions and problems, not a standardized instrument" (Moss 2008 as cited in Lyon 2013b, p. 2). Assessment in this view serves a formative function providing or facilitating learning opportunities. In the case of assessing and teaching linguistically and culturally diverse students, linguistically and culturally valid assessment practices should provide opportunities for these students to access the content and demonstrate their knowledge. Language functions play a central role as they determine the linguistic performance that students are expected to carry out in direct relation to the content and skills being assessed. Hence, language serves as an anchor for the science teachers to design assessment practices and scaffolds embedded in the assessment.

In agreement with Lyon (2013b), we have argued that language should no longer be viewed as an interference with the content knowledge assessed on content assessments. Rather, language is integral to the content being assessed. To allow students to access language on assessments, we have suggested that teachers provide opportunities for ELLs to respond to the assessment in multiple ways, including writing, speaking, and drawing instead of multiple-choice format only. Language of the assessment is not to be simplified or modified drastically, but scaffolds are to be used to allow ELLs to access the language of the assessment and demonstrate their knowledge.

We acknowledge that integrating fair and valid assessment practices in science classrooms is not easy both from a measurement and instruction perspective. Attending to language is often viewed as an inherently conflicted practice in assessment, where language is viewed as a source of construct irrelevant variance that interferes with the content knowledge being assessed. While some have taken this view to argue that language should be modified or simplified on content assessments (Abedi et al. 2004), others maintained that language of the content can be a resource for learning content and facilitating demonstration of content knowledge. Similarly, from an instructional perspective, language may be viewed as "a technical issue to control" instead of an integral part of content instruction and the act of assessment (Lyon 2013b, p. 8). Most teachers might feel that language demands need to be removed altogether to provide students with access to content. In fact, the three science teachers participating in the study (Lyon 2013b) straddled between reducing the language demands and scaffolding language use to enable the linguistic minority students to access the assessment and to demonstrate scientific understanding. However, over the course of the study, their assessment expertise evolved into an understanding that they in fact do not have to remove the language demands. The teachers recognized that they could help linguistic minority students to communicate scientific knowledge and understanding by letting them use language in various ways (i.e., writing, listening, speaking, and visual tasks). Another tension in instructional practice emerges from viewing assessment as a form of assigning grades to conceptual understanding of the content (Lyon 2013b). That is, the idea that assessment could serve for learning the language is hard to reconcile mostly because science teachers might not routinely look for language use in students' responses to the assessment.

These tensions are natural, especially when teaching and assessment do not work towards the common goal of learning through similar tools and methods. While in theory the two aim to facilitate learning, the tools and ways in which they go about that goal might not be directly conducive to learning. To us, one tool that unites teaching and assessment is language. Through encouraging appropriate, science-relevant language use, both teaching and assessment could mobilize the linguistic and non-linguistic resources all students, but especially linguistic minority students, bring to the learning environment.

Attention to language calls for iterative and systematic exposure and practice on how language functions in various discipline-specific discourse practices (e.g., making an argument based on evidence). The need for systematic orientation to language applies to both teacher education and professional development settings. Lyon (2013b) suggests that teacher educators showcase (via videos or examples of teachers assessing students) student written work and orient future or current teachers to understanding how language functions. The main emphasis here should be on helping pre-service and in-service teachers to balance reducing the language demands and providing opportunities for students to use discipline-specific language so they can access the content, develop, and demonstrate academic literacy.

The NGSS has already emphasized both enacting and assessing language-intensive practices in the classrooms. Science teachers are challenged by this goal, as they are not routinely trained and charged to address language as part of content instruction and assessment. Further, recent high stakes assessments present both a challenge and an opportunity for science teachers to address language in their classes. It is a challenge because these assessments are increasingly more linguistically demanding. However, this challenge provides an opportunity, because teachers can integrate linguistically and culturally responsive assessment practices in their classroom assessments and teaching in multiple ways that allow diverse learners to access and use language. To seize this opportunity, classroom assessments should be seen as powerful formative tools to leverage the opportunity for students to learn the language and content of science.

To conclude, we would like to highlight how critical it is for teacher education programs to provide strategies for science teachers suggesting how to support all students, including linguistically and culturally diverse students, in learning and using the scientific discourse effectively. It is also important for science teachers to view language as an integral part of the multifaceted scientific knowledge that all students need to develop. Moreover, teachers need to provide multiple opportunities for students to practice using the scientific discourse in the classroom.

Classroom assessments are integral parts of the learning process in the science classroom as they can provide valuable information about what all students know and can do in science, and they can model the kind of performances that are central to science. Consequently, teacher education programs should also focus on working with teachers on how to develop linguistically and culturally valid science classroom assessments. This means developing assessments that provide all students an opportunity to demonstrate what they know and can do. So, instead of simplifying the language in the assessments, teachers should provide different types of support or scaffolds so all students can access the science content and use language to demonstrate their science knowledge.

Acknowledgement We would like to thank Dr. Vinetha Belur and Joe Ciofalo for providing research support for the content presented in this chapter. We would also like to thank Dr. Michael Kane, Dr. Jamie Mikeska, Dr. Hui Jin, and Dr. Lei Liu for their constructive feedback to earlier versions of the chapter.

References

Abedi, J., Hofstetter, C. H., & Lord, C. (2004). Assessment accommodations for English language learners: Implications for policy-based empirical research. *Review of Educational Research, 74*(1), 1–28.

Abedi, J., & Lord, C. (2001). The language factor in mathematics tests. *Applied Measurement in Education, 14*(3), 219–234.

Aud, S., Hussar, W., Kena, G., Bianco, K., Frohlich, L., Kemp, J., Tahan, K. (2011). *The Condition of Education 2011 (NCES 2011–033).* U.S. Department of Education, National Center for Education Statistics. Washington, DC: U.S. Government Printing Office.

Avenia-Tapper, B., & Llosa, L. (2015). Construct relevant or irrelevant? The role of linguistic complexity in the assessment of English language learners' science knowledge. *Educational Assessment, 20*(2), 95–111.

Bailey, A. L. (2007). *The language demands of school: Putting academic English to the test.* New Haven and London: Yale University Press.

Basterra, M. (2011). Cognition, culture, language, and assessment: How to select culturally valid assessments in the classroom. In M. Basterra, E. Trumbull, &

G. Solano-Flores (Eds.), *Cultural validity in assessment: Addressing linguistic and cultural diversity* (pp. 254–274). New York, NY: Routledge.

Basterra, M., Trumbull, E., & Solano-Flores, G. (Eds.) (2011). *Cultural validity in assessment: Addressing linguistic and cultural diversity*. New York, NY: Routledge.

Buxton, C. A., Allexsaht-Snider, M., Suriel, R., Kayumova, S., Choi, Y. J., Bouton, B., & Baker, M. (2013). Using educative assessments to support science teaching for middle school English-language learners. *Journal of Science Teacher Education, 24(2)*, 347–366.

Common Core State Standards Initiative. (2010). Common core state standards for English language arts & Literacy in history/social studies, science, and technical subjects. Washington, DC: National Governors Association Center for Best Practices and the Council of Chief State School Officers.

Dutro, S., & Kinsella, K. (2010). English language development: Issues and implementation at grades 6 through 12. In California Department of Education (Eds.) *Improving education for English learners: Research-based approaches* (pp. 151–208). Sacramento, CA: California Department of Education Press.

Fang, Z. (2006). The language demands of science reading in middle school. *International Journal of Science Education, 28*(5), 491–520.

Fradd, S. H., & Lee, O. (1999). Teachers' roles in promoting science inquiry with students from diverse language backgrounds. *Educational Researcher, 28(6)*, 14–42.

Fusco, D., & Barton, A. C. (2001). Representing student achievements in science. *Journal of Research in Science Teaching, 38*(3), 337–354.

Hakuta, K. (2014). Assessment of content and language in light of the new standards: Challenges and opportunities for English language learners. *The Journal of Negro Education, 83*(4), 433–441.

Kane, M. T. (1992). An argument-based approach to validity. *Psychological Bulletin, 112*(3), 527.

Kane, M. T. (2001). Current concerns in validity theory. *Journal of Educational Measurement, 38*(4), 319–342.

Kane, M. T. (2006). Validation. *Educational Measurement, 4*(2), 17–64.

Kelly-Jackson, C., & Delacruz, S. (2014). Using visual literacy to teach science academic language: Experiences from three preservice teachers. *Action in Teacher Education, 36*(3), 192–210.

Kopriva, R., Winter, P., Triscari, R., Carr, T. G., Cameron, C., & Gabel, D. (2013). *Assessing the knowledge, skills, and abilities of ELs, selected SWDs, and controls on challenging high school science content: Results from randomized trials of ONPAR and technology-enhanced traditional end-of-course biology and chemistry tests*. Retrieved from http://onpar.us/research/reports.html

Lee, O. (2003). Equity for culturally and linguistically diverse students in science education: A research agenda. *Teachers College Record, 105*, 465–489.

Lee, O., Santau, A., & Maerten-Rivera, J. (2011). Science and literacy assessments with English language learners. In M. Basterra, E. Trumbull, & G. Solano-Flores (Eds.), *Cultural validity in assessment: Addressing linguistic and cultural diversity* (pp. 254–274). New York, NY: Routledge.

Lee, O., Quinn, H., & Valdés, G. (2013). Science and language for English language learners: Language demands and opportunities in relation to Next Generation Science Standards. *Educational Researcher, 42*(4), 423–433.

Lee, O., & Fradd, S. H. (1998). Science for all, including students from non-English language backgrounds. *Educational Researcher, 27*(4), 12–21.

Lemke, J. L. (1990). *Talking science: Language, learning, and values*. Norwood, NJ: Ablex.

Lyon, E. G., Bunch, G. C., & Shaw, J. M. (2012). Language demands of an inquiry based science performance assessment: Classroom challenges and opportunities for English learners. *Science Education, 96*(4), 631–651.

Lyon, E. G. (2013a). Conceptualizing and exemplifying science teachers' assessment expertise. *International Journal of Science Education, 35*(7), 1208–1229.

Lyon, E. G. (2013b). What about language while equitably assessing science? Case studies of preservice teachers' evolving expertise. *Teaching and Teacher Education, 32*, 1–11.

Martiniello, M. (2008). Language and the performance of English-language learners in math word problems. *Harvard Educational Review, 78*(2), 333–368.

McNeill, K. L., & Pimentel, D. S. (2010). Scientific discourse in three urban classrooms: The role of the teacher in engaging high school students in argumentation. *Science Education, 94*(2), 203–229.

Messick, S. (1989). Meaning and values in test validation: The science and ethics of assessment. Educational Researcher, 18, 5–11.

Mislevy, R. J., & Durán, R. P. (2014). A sociocognitive perspective on assessing EL students in the age of common core and next generation science standards. *TESOL Quarterly, 48*(3), 560–585.

Moje, E. B., Collazo, T., Carrillo, R., & Marx, R. W. (2001). 'Maestro, what is 'quality'?' Language, literacy, and discourse in project-based science. *Journal of Research in Science Teaching, 38*(4), 469–498.

Moss, P. A. (2008). Sociocultural implications for assessment I: classroom assessment. In A. M. Moss, D. C. Pullin, J. P. Lee, E. H. Haertel, & L. J. Young (Eds.), *Assessment, equity and opportunity to learn* (pp. 222–258). Cambridge: Cambridge University Press.

NGSS Lead States. (2013). *Next generation science standards*. Washington, DC: National Academies Press.

Noble, T., Rosebery, A., Suarez, C., Warren, B., & O'Connor, M. C. (2014). Science assessments and English language learners: Validity evidence based on response processes. *Applied Measurement in Education, 27*(4), 248–260.

Pennock-Roman, M., & Rivera, C. (2011). Mean effects of test accommodations for ells and non-ells: A meta-analysis of experimental studies. *Educational Measurement: Issues and Practice, 30*(3), 10–28.

Pullin, D. C., & Haertel, E. H. (2008). Assessment, equity, and opportunity to learn. In A. M. Moss, D. C. Pullin, J. P. Lee, E. H. Haertel, & L. J. Young (Eds.), *Assessment, equity and opportunity to learn* (pp. 17–40). Cambridge, United Kingdom: Cambridge University Press.

Schleppegrell, M. J. (2001). Linguistic features of the language of schooling. *Linguistics and Education, 12*(4), 431–459.

Siegel, M. A. (2007). Striving for equitable classroom assessments for linguistic minorities: Strategies for and effects of revising life science items. *Journal of Research in Science Teaching, 44*(6), 864–881.

Siegel, M. A. (2014). Developing preservice teachers' expertise in equitable assessment for English learners. *Journal of Science Teacher Education, 25*(3), 289–308.

Solano-Flores, G., & Nelson-Barber, S. (2001). On the cultural validity of science assessments. *Journal of Research in Science Teaching, 38*(5), 553–573.

Stoddart, T., Solis, J., Tolbert, S., & Bravo, M. (2010). Effective science teaching for English language learners (ESTELL). In D. Sunal, & C. Sunal (Eds.), *Teaching science with Hispanic ELLs in K-16 classrooms* (pp. 151–181). Albany, NY: Information Age Publishing.

Trumbull, E., & Solano-Flores, G. (2011). The role of language in assessment. In M. Basterra, E. Trumbull, & G. Solano-Flores (Eds.), *Cultural validity in assessment: Addressing linguistic and cultural diversity* (pp. 22–46). New York, NY: Routledge.

Turkan, S., & Liu, O. L. (2012). Differential performance by English language learners on an inquiry-based science assessment. *International Journal of Science Education, 34*(15), 2343–2369.

Warren, B., Ballenger, C., Ogonowski, M., Rosebery, A. S., & Hudicourt-Barnes, J. (2001). Rethinking diversity in learning science: The logic of everyday sensemaking. *Journal of Research in Science Teaching, 38*(5), 529–552.

Wolf, M. K., & Leon, S. (2009). An investigation of the language demands in content assessments for English language learners. *Educational Assessment, 14*(3–4), 139–159.

Young, J. W. (2009). A framework for test validity research on content assessments taken by English language learners. *Educational Assessment, 14*(3–4), 122–138.

Sultan Turkan is a Research Scientist in the Student and Teacher Research Center at Educational Testing Service in Princeton, New Jersey. Her research focuses on culturally linguistically responsive teaching of disciplinary literacy for bilinguals/multilinguals and equitable assessment of these students. Dr. Turkan is the lead author of papers that provide alternative approaches to addressing important issues in the education of bilingual/multilingual students. An important contribution of her work is the concept of *disciplinary linguistic knowledge*, key to specifying the sets of skills content teachers need to develop in order for their teaching to be sensitive to the needs of bilinguals/multilingual students. Her work appeared in *Teachers College Record, Educational Researcher, Theory into Practice,* and *Urban Education.*

Alexis A. Lopez is a Research Scientist in the Center for English Language Learning and Assessment at Educational Testing Service in Princeton, New Jersey. His research focuses on the assessment of English language proficiency and the assessment of content knowledge for K-12 English Learners. He is currently involved in several research projects focusing on developing formative assessments and flexible bilingual content assessments for English learners. He has published some of his work in *Educational Assessment,* and the *Encyclopedia of Language and Education, Language Testing and Assessment.*

9

Practical Language Learning Strategies that Increase Science Learning and Engagement

Ana Lado and Adrienne Wright

One day, Ana was circulating around the class monitoring the summaries being written by small groups of English Language Learners (ELLs). She found that their writing was confusing. By questioning them, it became clear that they had misinterpreted the word "grounds" as in the grounds for a decision with the concept of "grounds" as in earth and fields. She intervened by implementing a language teaching strategy in which students are taught key science words by associating them with actions and using them in sentences, such as making a motion for bouncing while saying "The soccer ball bounced on the ground." and making another motion for the other meaning while saying "The data collected about plant growth became the grounds for the watering schedule." Then the ELLs returned to the original assignment, the summaries, with greater success.

A. Lado (✉)
Department of Education, Marymount University, Arlington, USA
e-mail: alado@marymount.edu

A. Wright
Department of Biology, Arlington County Schools, Virginia, USA
e-mail: adrienne.wright@apsva.us; ettemazme@aol.com

© The Author(s) 2017

L.C. de Oliveira, K. Campbell Wilcox (eds.), *Teaching Science to English Language Learners*, DOI 10.1007/978-3-319-53594-4_9

When English Language Learners (ELLs) are learning science, they are often overwhelmed because they are experiencing cognitive overload given that they are learning a new language along with science (Cummins 1981). It is overwhelming to process new language sounds, words, sentences, and paragraph structures while at the same time processing new science content. Many ELLs are confused by incongruities between the language of expository text and the teacher's oral instructional language. Even when the differences are understood, they are confused by being expected to seamlessly transition back and forth between oral and written language. In addition, science texts often assume background knowledge that some ELLs may not yet have. Faced with an overload, typically ELLs become reticent to speak and are less engaged in learning (Mohr and Mohr 2007). In order to avoid communication breakdowns, teachers must adjust to ELLs' English language proficiency levels in the science classroom.

In this chapter, we provide examples from a High Intensity Language Training Extension program (HILTEX). HILTEX serves ELLs from diverse linguistic and cultural backgrounds. The students in these examples are in WIDA levels 3 and 4.[1] In the school system from which the scenes are taken, students in these levels attend a two-period reading/language arts block class taught by an endorsed ESL teacher and mathematics, science, or grade level teacher according to whether the class is in mathematics, science, or an elective. They are expected to take and pass statewide standardized subject matter tests.

In our science classes with ELLs, we reduce cognitive overload by balancing a focus on language usage with a focus on science knowledge. We make use of and create redundancies and repetition among the

---

[1] WIDA stands for the World-Class Instructional Design and Assessment, a consortium of states dedicated to the development and implementation of English language proficiency standards and tests for ELLs. Levels 3 and 4 fall in the middle of the 6 WIDA levels, so these students are considered to be intermediate ELLs who are developing and expanding beyond the beginner level. The performance descriptors for grades 9–12 can be found on the WIDA website www.wida.us 2012 ELD standards page.

building of literacy, oral, and science skills. When materials, teacher talk, and student comprehension work together, there is a synchronicity which creates opportunities for exponential student learning. The teacher must create a warm and approachable relationship that teaches English together with science continuously. Our use of Communicative Language Teaching (CLT) strategies permeates our interactions with ELLs.

Adrienne stands by the classroom door and greets each student by name as they enter the classroom every day. They adore this simple welcoming gesture, to the point that many of them stand there with her and greet others on their way in. Their participation in this simple greeting leads to a camaraderie during lessons. They learn to imitate the way she approaches texts. They learn to laugh at mistakes. They know that she gives them the space, time, and models to keep learning.

We have implemented the types of conversational uptakes and connections that have been written about by Perez (1996). Perez describes ways to respond to ELLs that respect the ELL student and provide linguistic scaffolds that foster more and better discussion of academic topics. Mohr and Mohr (2007) increased ELLs' participation by matching their response according to the type of responses ELLs produce, regardless of whether the ELLs' responses were silence, undecipherable, or ungrammatical. ELLs need teachers to respond to them using strategies that validate their efforts and provide appropriate language models to guide them to successful participation. Otherwise ELLs feel overwhelmed, invisible, out of place, and are hesitant to ask for help or participate.

The aim in this chapter is to show CLT strategies with science content and academic language usage. These strategies free the teacher from extra explanations of the readings and help ELLs become independent users of English while building science foundational skills. The CLT strategies we use are based on three theoretical strands, the principles of CLT (Nation 2008), foundational concepts in teaching ELLs academic content (Nutta et al. 2011; Echevarria and Graves 2003; Cummins 1981), and Labov's conversational analysis (1972). After outlining these, we will provide examples of ways the CLT strategies on which they are based are implemented to increase ELL success.

We use CLT strategies in a three-step (1/ whole class 2/ small groups 3/ whole class) routine. It has resulted in enthusiastic and confident participation in science class. The ELLs recognize that what they once considered overwhelming has become a manageable task. They learn ways to develop their English language competence at the same time that they learn ways to access and study content knowledge. In each step, the whole class introduction, the small group work, and the whole class review, we apply strategies that flow from foundational concepts in CLT.

Theoretical Foundations

The foundations of our teaching are contained in Labov's research in conversational analysis (1972), specifically the notion that successful conversations are not the result of the content of the story but of synchronicities among the participants. He coined the term *tellability* in describing successful conversations and we will apply it to successful instruction with ELLs. One way to create synchronicity is to implement a balanced program of CLT, i.e., to include comprehension, expression, language study, and fluency activities that are simplified (Nation 2008). Simplification within teaching content with ELLs is described by Cummins's (1981) as the intersection of appropriately demanding cognitive content and context rich language. When we teach less content we have more success in language, less is more.

Tellability

Labov's work with conversational analysis helps ELLs by talking through the highlighted vocabulary in the text, and also adjusting to the needs of ELLs as they arise. The purpose is to create synchronicity of the teacher's oral instruction with the students and the content material. It involves planned whole group instruction strategies followed by small group activities. It involves spending more time on foundational science and language understandings and less time on extraneous concepts and details. It does not involve much more preparation of materials but it does involve more linguistic guidance.

Labov coined the term to describe the need for the speaker to adapt to the listeners and the context. In our science classrooms, this means that we examine not only the difficulty in the written text, but also the oral instructional conversations and additionally their alignment to the written science materials and content.

Labov's work allows us to shift from the traditional focus on the difficulty of the text alone, because a text's readability is calculated on formulas based on vocabulary and sentence length. This is devoid of consideration of key variables, such as the educational context and linguistic abilities of the students. In reality, determining the difficulty of a text must include instructional variables and student characteristics (see Harris and Hodges 1995). Our focus is on the difficulty of the entirety of the instructional context which includes the students, the materials, and the oral language used by the teacher. Since our students are ELLs, we must adjust the material given their English abilities. Since the concept of tellability requires us to examine the entirety of the situation and not just the text, we reduce the overload for ELLs by reducing the difficulty of the materials. We also reduce the overload by reducing the difficulty encountered with our instruction by using CLT strategies.

Matching oral instruction to passages in the written text will lessen the linguistic burden for ELLs (Cummins 1981). It allows for ELLs to engage in deep processing of content and language together, because they are not learning two styles of English at once, the English of the written science texts and the English of the instructional conversation.

Less Is More

One of the major theoretical undercurrents we use is the notion of less is more, sometimes referred to as simplification (Nation 2008; Cummins 1981). It refers to the fact that in teaching English as a second language, we modify the English so that ELLs have the opportunity to integrate the recognition of sounds, symbols, words, sentences, and discourse structures. ELLs must master the simpler and fundamental structures on which the more complex English ones are based. Extraneous language is removed or ignored and ELLs are shown models of clear and concise paragraphs with the essential vocabulary.

We simplify language and content (Nunley 2004). ELLs need the science concepts to be presented in concise, simplified, clear language written in consistent ways (Nation and Deweerdt 2001; DaSilva Iddings et al. 2009; August and Gray 2010). The Science Writing Heuristic (SWH) uses a few key questions that result in an appropriate amount of simplification (Keys et al 1999). We use it to ask about the information that ELLs need to know, the language that they find difficult in the text, and the academic words that represent key concepts. As an example of focusing on information that they need to know, we can help them understand text features that distinguish what is important from what is less so, such as tables of content, highlighted words, and glossaries. As example of accessing difficult text and key concepts, we help them learn to define words using a specific sentence pattern. ELLs benefit from explicit modeling of ways to use text features to find important information, ignore extraneous language and content, and process key words and difficult text. Modeling is a foundational concept in CLT.

Explicit Modeling

Modeling is a part of all four strands of a balanced CLT program. We use a range of modeling strategies for comprehending English, learning English expression, studying vocabulary and grammar, and developing fluency (Herrell and Jordan 2012; Nation 2008; Burke et al. 2005; Linik 2004). For comprehension we use visuals, gestures, realia,[2] and other scaffolds during interactions. For expression, we model routines of scientific ways of thinking, giving students tasks that practice the language pattern used, and expecting them to reproduce the teacher's examples orally and in writing. We model ways to study academic English, since the ELLs in HILTEX have mastered the everyday English words (found on the General Service List[3]) but need help with academic English and the English particular to science.

[2] In CLT "realia" is the term used for manipulatives, models, and miniature objects.

[3] The General Service List (GSL) contains the 2000 most frequent words of English (Nation 2008) and is found on http://www.newgeneralservicelist.org/.

Studying academic words is important for understanding the overall text structure. The Academic Word List (AWL)[4] are words that appear with most frequency across all academic texts. ELLs need these words for science as well as other academic subjects. We teach academic words in various ways. We use the ELLs' background knowledge and spoken ability to connect them to general English words. We engage the ELLs in going back and forth between oral tasks and written ones. We deliberately engage in word study because 91% of the AWL are words of French, Latin, or Greek origin (Nation 2008).

We engage ELLs in guided conversations that connect everyday action verbs to academic ones. For example, we teach ELLs to connect general English words to academic equivalents, such as "tell" and its equivalent "describe," and "retell" and its equivalent "summarize." We also teach the nominalizations so commonly used in science texts (e.g., germination, pollution). We begin guiding students with physical actions. ELLs can act "plant (v)" and from there learn the meaning of germinate and germination and can act "toss" as a way to introduce the meaning of pollute and pollution. Explicit modeling continues when we teach them to the use the new vocabulary in sentences. For example, "We tossed the object" and "Discarding the waste contributed to water pollution."

The different types of vocabulary found in science texts imply the use of different types of strategies.

Technical words are different from academic ones because they are specific to academic topics. Just as with academic words, we model studying word parts, and defining them in ways that relate to its core meaning (Nation 2008). For example, we define them within a graphic organizer of the topic.

Simplification does not mean watering down the essential curriculum vocabulary. Adrienne gives this example of the way she addresses the need to teach ELLs technical terms. The SOL in Biology covers the

---

[4] The Academic Word List (AWL) (Nation 2008) contains 570 word families that appear with great frequency in a broad range of academic texts. The list does not include the most frequent 2000 words of English (the General Service List), thus making it specific to academic contexts. It can be accessed on https://en.wikipedia.org/wiki/Academic_Word_List.

terms exothermic and endothermic. We stop and examine the words by answering questions, such as "What does -THERMic make you think of? by showing them a thermometer and introducing the term thermostat. They may be unfamiliar with the term, but when the control on the wall in their house or apartment is described, when they are shown a picture of one, and when I point out the one in the classroom, most of them know what it is. Then, we introduce ENDO- and EXO- connecting EXO- to exit and EXOskeleton. I will pass around a grasshopper exoskeleton as an example of a skeleton on the outside of the body, and then ask, "Where is your skeleton?" They reply, "On the inside" and continue on to a description of an ENDOskeleton, "an internal skeleton." Perhaps they contribute that they see a connection between ENDO and entrance. Then we compare and contrast ENDO- and EXO- thermic reactions.

The deliberate teaching of vocabulary helps ELLs who may not have basic study skills and can learn from teachers who narrate what they are thinking and doing, and ways that they study words. It models an academic way of thinking but also clarifies obscure concepts for ELLs. In short, explicit modeling with ELLs is unlike the talk in regular classes, in that the teacher models constantly, even ways to complete assignments. Modeling is purposeful and is the first step to guiding ELLs to engage in class.

To summarize our implementation of theory, teaching ELLs requires teaching strategies that provide more explicit modeling and guiding, we teach more fundamentals and less extraneous details, and connect oral instructional language to written text language. In order to implement the theory the classes with ELLs follow a three-step procedure which creates flow among the teacher's instruction, the students' participation, and the instructional situation in which English and science are taught simultaneously.

We now move to discussion of the three-step process of implementing the theory. The first step in our classroom routine is with the whole class. CLT strategies are used to scaffold their learning from texts used with the mainstream. The second step involves small groups that work cooperatively to personalize and individualize the material. The teacher circulates among the groups. The third step brings the class together again to review, rehearse, and build fluency.

Three-Step Classroom Implementation

The three-step routine of whole-group/small-group/whole group has exponentially increased the successful participation of ELLs. The teacher begins with whole class instruction and models key concepts and skills. The ELLs then move to small groups in which the teacher is "*doing the laps*" (circulating from small group to small group multiple times) and continues to bridge the gap for students in language and science and gradually decreases the need for guidance as they increasingly improve their processing of the language of science. The goal is to require ELLs to participate by guiding them in the English needed to show their developing science understanding. The following visual recreates this whole class to small group sequence (Fig. 9.1).

Step 1: Introduce the Language and Content with the Whole Classroom

Vignette. *Adrienne surveyed her HILTEX students, asking them about ways they need instruction. Most respond that explicit, whole group, direct instruction really helps them, but with a caveat. They also said that they find themselves lost in classes without adapted explanations of the directions or content. When they find a teacher's talk too fast to grasp, they are reticent to stop a teacher to ask for clarification. They wanted the teacher to speak slowly and enunciate carefully, pause frequently to check for understanding, and stop to teach academic vocabulary. But, they also wanted her to deliberately teach them study skills that they could apply to all their difficult content classes. This is when she began to set aside time for teaching them ways to study academic words found in all their texts and to study word parts that unlock word meaning.*

During whole-class direct instruction it is important to have more of a discussion than a one-sided lecture. While we use simplified text, we also challenge them to learn from authentic science texts that they will encounter in statewide standardized tests. Specific academic language and strategies for accessing authentic texts must be taught (Lee 2005). The teacher uses Think Aloud and stops to ask a lot of probing, open-ended questions to ferret out misunderstandings.

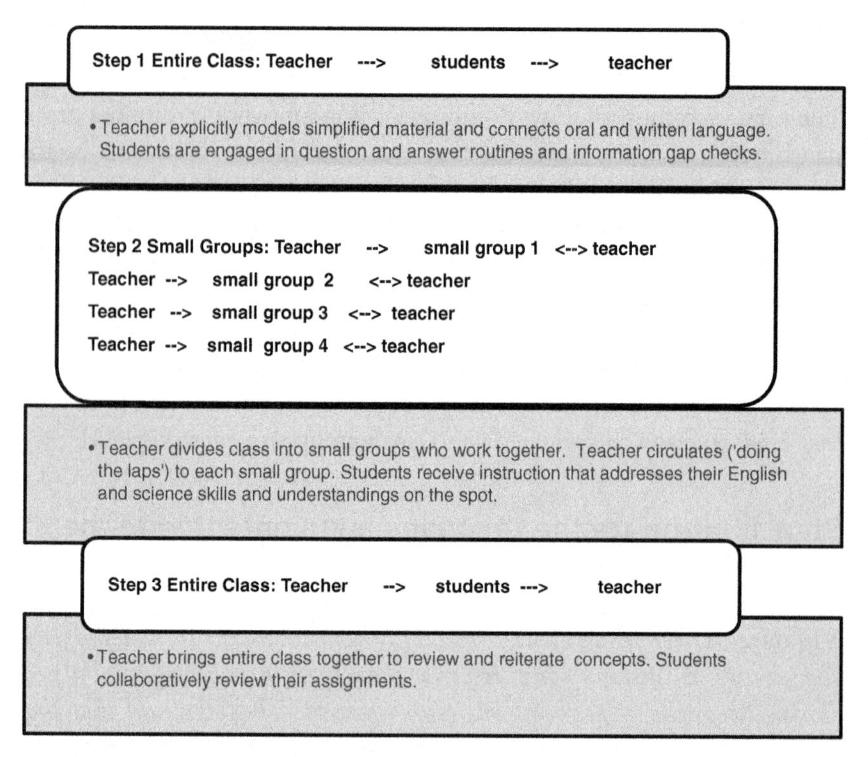

Fig. 9.1 Three steps: The scheduling of three times for simultaneously validating and modeling participation with ELLs

Think Aloud with text. Think Aloud refers to a strategy in which teachers model a reasoning process in order to guide students. When introducing a new topic or activity, such as a science laboratory protocol, the teacher projects the same document on the screen as the one used in a regular, mainstream class. This saves the teacher time, because the teacher does not need to create a new protocol for the ELLs, but instead teaches it differently. They explicitly go over the existing one designed for native English speakers, yet focusing on demonstrating thinking about the essential vocabulary and skills. In addition, ELLs get their own hard copy to mark.

ELLs mark this copy while the teacher narrates, reads through the protocol. As they progress through it, the teacher simplifies and clarifies complex sentences, examines both academic and content vocabulary, and restates and summarizes the parts of the protocol, all while making notations on the document to clarify English and science concepts. During this reading, the teacher implements the typical strategies of adding pictures and realia for unfamiliar vocabulary and concepts. The ELLs see the teacher thinking aloud and making notations. They copy this thinking and these notes onto their own protocol. The ELLs acquire practice in examining text at the level expected in the mainstream classes and on standardized tests, and benefit from the experience of learning how to unpack meaning in texts that would otherwise be inaccessible to them without this exposition of the teacher's Think Aloud.

The teacher asks the ELLs to restate during whole-group direct instruction to ensure that they understand the directions, purpose, and outcome of the text-based learning activity. By explicitly scrutinizing text together with the teacher, ELLs gain confidence with the complicated academic and content English used in science. This thinking aloud with the text is also useful for teaching the vocabulary and the sentence structures of texts.

We demonstrate the need to stop to think consciously along the way to examine the vocabulary patterns, the discourse patterns, and predict meaning based on these. For example, we conduct word study of word roots, affixes, synonyms/antonyms, and analogies. This word study enables us to use basically the same materials as the regular classes. This is important because the end-of-course tests are for all students, including the ELLs. The word study during the year has prepared them for the words that they will likely see on the test, and prepares them with a procedure for ways to decipher an unfamiliar word or a complex sentence they may encounter.

Questions and follow-up responses. During this step with the whole class, the teacher engages in explicit modeling of the patterns of English by implementing its use until the students understand routines and patterns of academic English (Perez 1996; Mohr and Mohr 2007; Dreher and Gray 2009; August and Gray 2010). Often predictable question and answer routines, question and answer pairs, and sentence stems are used to guide ELLs to learn academic English. Teachers must model and provide guided practice with the typical language used during instruction.

Dreher and Gray (2009) show ways to have ELLs recognize, compare, and contrast discourse linguistic structures and produce them.

One example of the ways teachers use linguistics is when they model the structure of the answer or response. Teachers can teach them to always begin to answer the question,

"How do you know?" with "I know___ because ___."
This pattern can also be used in answer to the question,
"What do you think and why?" "I think ____ because ___."

A heuristic we consulted for teaching ELLs to examine a written text and eventually guide them to write a report is the Science Writing Heuristic (SWH) (Keys et al. 1999). The SWH questions include the following:

1. What are my questions?
2. What do I do?
3. What can I see?
4. What can I claim?
5. How do I know? Why am I making these claims? (evidence)
6. How do my ideas compare with other ideas?
7. How have my ideas changed? (Burke et al. 2005)

Teaching ELLs routines through patterned questions and answers helps them build confidence in answering. We use the same questions and rehearse answers as many times as students need them to feel sure about their ability to answer. We require ELLs to answer. They must also know that there is no penalty for making mistakes while learning. Some of them are so shy at first that speaking in class is very hard for them, even if they know what to say. Teachers provide them the scaffolds and give them lots of practice. The camaraderie of a supportive learning environment leads to a feeling of succeeding together.

Step 2: Small Group Targeted Instruction

Vignette. *The English speakers in class were easily engaged in a student-directed activity after a whole class lecture. But the ELLs seemed overwhelmed with the task of writing a summary. I arrived at one particular* *small group, and noticed that they had produced a confusing and disjointed summary. As I probed for the misconceptions, I realized that they had confused "psycho" with the idea of "cycle" during the lecture at the beginning of class (step one). This initial confusion probably prevented them from making key connections among the concepts all along and therefore they had missed some of what I had been teaching. I intervened. First, we reviewed the notes we had made on our protocol from step one. Second, we put some of the key vocabulary words onto a graphic organizer. Third, we quickly reviewed the sentence stems used to write each part of a summary paragraph. Then I left them in order to continue onto another small group. I left with confidence that their next attempt would have clarity.*

The vignette illustrates the need to plan targeted instruction for the ELLs' abilities in language and content. The technique of the teacher circulating around the room, which we call "doing the laps," provides ELLs with opportunities to orally demonstrate their understanding and does two things: first, it enables the teacher to immediately assess what they know and do not know and offer immediate feedback, and second, it gives the ELLs practice with speaking about content in a way that requires synthesis of newly acquired content as well as using the English language in a way that challenges them appropriately. It is important that this be formative rather than summative, enabling them to practice and receive feedback without penalty.

The type of direct instruction needed by ELLs depends on their abilities, and this is the reason that it is so successful to follow whole group instruction with small group instruction and after the opportunities for ELLs to engage in small groups, return to the large class again. Small group work increases opportunities for ELLs to learn through social means from each other, but it also gives the teacher a platform for reaching ELLs with different backgrounds and abilities.

For example, recall that there are three types of vocabulary being taught in a science class with ELLs, general English words, ubiquitous academic words, and technical topic-specific science words. ELLs who are at different levels of English proficiency require different teaching responses depending on whether they have productive capabilities with each of these (Swain 1995; Cloud et al. 2009). The first task in circulating around the small groups is to assess their understandings.

ELLs orally demonstrate their understanding with such questions as, How are these two similar? How are the first two different?

They are required to answer, and based on their answers, direct instruction is adjusted accordingly. For example, some ELLs are stronger in receptive English. They understand what they hear or read, but have trouble verbalizing it. These students are given models to repeat and patterns to practice. Other ELLs are stronger in speech. They overuse expressions such as get into, uptake, put into, and get it. I give these students direct instruction in substituting these phrases with the collegiate (academic word list) equivalent word, such as get into—examine, uptake—employ, put into—apply, and get it—utilize.

While some of these adjustments can be planned for ahead of time, it is more effective to use them on the spot while students complete tasks where they are trying to reflect on, clarify, and produce a summary or paraphrase. Groups are given additional tasks to reach the challenge of assimilating the concepts. They may be given a model summary or definition that they must use for their assignment. One group is given a set of questions, another only the sentence stems, and still another group the reference in the text to incorporate. These on-the-spot adjustments allow students to feel a part of the discovery and learning.

ELLs can deeply process new information through routinely using the same types of small group assignments, such as information gap checks and pre-determined question and answer pairs. The purpose of the small group activities is to work together on tasks that increase their essential understandings, knowledge, and skills. During step two, ELLs know that they will not be asked to demonstrate mastery. The idea in step two is to have them engage by working with questions that they still have trouble answering and information that they are still learning. This is a key part of the second step and the group work. They know that they have not yet mastered the necessary content, cannot yet effectively express it in English, or both. They will learn it with help from their group, the teacher, and later, the whole class. The teacher regularly uses the same strategies to challenge them in small groups. They must work together to complete the task or solve the problem. In doing so, they master the English skills used in expressing the information and the solution. The examples that follow involve conversational uptakes and information gap checks.

Conversation uptakes and verbal responses. During small group work when "doing the laps," students are asked probing questions. For example, in a unit about cells, the teacher would identify essential understandings, knowledge and skills for each topic, and formulate several questions for each one. If the essential knowledge is *describe how the selective permeability of the cell membrane affects the life of a cell,* then the teacher could formulate these questions:

What does selective permeability mean?
Why does a cell membrane need to be selectively permeable?
What components of the cell membrane contribute to its selective permeability?
How do those components function in selective permeability?

When students are working in small groups, the teacher can move from table to table asking these questions, to which students would respond verbally. These questions focus the students on what they should be researching and discussing with their group in order to gain the essential knowledge. Most of them will try to write down the questions that you ask them. This is an excellent way for them to take responsibility for their learning and to learn the skill of transferring back and forth between oral and written language.

The ELLs get multiple opportunities to practice verbally answering the questions that were raised in the whole class lecture. The goal is to get the students to where they can answer the questions without referring to any written material (and have therefore committed it to memory), and in the meantime they are practicing their speaking and listening skills, as well as their reading skills while conducting their research with each other and with the teacher. Later, when required to demonstrate the essential knowledge on a summative assessment, they have already parsed what questions need to be considered to fully address it, practiced answering them verbally, received feedback, and made any necessary adjustments to their understanding and/or verbal response. At this point they would simply write down what they had practiced speaking verbally, and they know it is correct and is effectively communicated, as they had already received that feedback earlier.

This question and answer verbal response routine is powerful because it puts them on the spot initially, and they will struggle while they are still learning the content. At first, they mostly cannot answer the questions. This creates a positive stress, and focuses them on what they do not understand or cannot answer, and they will actively seek to relieve the stress by searching for the information they are missing, or trying out different ways to verbally express themselves effectively. This stress will also help to create a lasting memory of their experience, and what they learned from it. They will practice multiple times with their group and the teacher, honing their understanding and responses each time.

Additionally, with guidance from the teacher, they quickly experience success when able to (with help and practice) answer a question to the teacher's satisfaction, and buoyed by this initial success, will be motivated to tackle the next question, and so on. The students must know that they can make mistakes without penalty during this process; otherwise they will experience negative stress that prevents them from learning. They are stretching themselves, venturing into unfamiliar territory, and need to feel supported while doing so. After a session like this, many students literally jump up and down with excitement, because they mastered a concept after struggling through discomfort and confusion, to emerge with a new understanding of both a content objective and how to clearly demonstrate it. In English! We adapt to the needs of ELLs for integrated lessons and interactive learning. They learn to study language and they learn to study science. They do this during whole class time but in small group instructional as well. In small groups they practice transferring information between written text and oral language styles. They also practice explicit study as well as indirect learning through small group interactions with each other and with the teacher, who is doing laps, moving from group to group for support during the activity.

An added benefit to this oral question and answer technique during the activity in small groups is that the interaction is entirely verbal and virtually instantaneous between teacher and student. What students understand and do not understand becomes immediately clear to both the teacher and the ELL, providing an opportunity to correct misunderstandings or fill in gaps on the spot. This prompt feedback is useful in that the ELL isn't waiting for feedback on an assignment they turned in a week ago long after such feedback would have been most useful. It accelerates the pace of learning. The teacher has a method for formative assessment that is highly effective without having to grade a stack of papers. The teacher could simply be checking off that the student has mastered the knowledge or skill when they have done so.

Information gap checks. We give small groups information gap checks which help them process key concepts and can be used to monitor mastery. Information gaps are activities in which some students have information that others do not have. An example might be giving different students paragraphs in which each has different information missing. They must work together to fill in their gaps. For example, during an activity investigating the effect on cells which are placed in different solutions, the ELLs would ask other students for information missing on each of their worksheets that other students have on their worksheets. For the following text, different students have different information in their texts and so one would ask the other "What is the effect?" and "What is plasmolysis?" while the other, "What is hypertonic?" The back and forth of information gap questions between students helps them understand science texts. "In this activity, we want to see the effect on a (cell) when placed in different (solutions). We expect that when a plant cell is placed in a (hypertonic) solution, the cell will (lose water and shrink). This is called (plasmolysis)." The teacher visiting each group provides the specific help needed for individuals to notice meaningful language patterns in the text, answer questions, and expected in the assignment.

Step 3: Collaborative Review

Vignette. *Adrienne impresses her students with her ability to list all 50 United States in alphabetical order in under a minute, due to having learned a song in the 5th grade (quite a long time ago!). I do this as an example of how powerful song/chanting can be when committing something to memory and I use it to begin an activity in which they must commit scientific facts to memory.*

Science requires the learning of lots of new concepts and information. ELLs need help in developing strategies to grasp the information. We do this with increasing activities that involve them in reinforcement, review, and processing information. We highlight information in the whole class instruction and afterwards, and reinforce it right away. Again, following an I-We-You pattern, the teacher follows up with a skill development activity that we do together. The teacher models it and they follow along as a whole group. Then, they will be given those skills to work on in small groups, and they practice while the teacher circulates around the room answering questions, and correcting misconceptions. Finally, the teacher will verbally quiz them (formatively) on the spot to check for understanding. This gives them the opportunity to listen to questions and speak their answers independently.

This is highly valuable as it provides immediate feedback for the ELL, and allows the teacher to quickly determine what they know. In mainstream classes, reading and writing is overemphasized, and ELLs often don't get the practice they need to listen, and explain and describe communicatively. In our class there is plenty of reading and writing, but also much listening and speaking. It is not a one-sided teacher talks, student listens, student reads and writes, but a two-sided, supported discussion involving talking, listening, reading, and writing. During the third step of the lesson, there is a collaborative review of assignments and material. Unlike in regular classes, in a HILTEX class the teacher models and reviews assignments together with the class.

Along the same lines as when introducing assignments and the text, the teacher demonstrates ways students are to stop and examine text. ELLs learn predictable ways to examine language patterns. For example, we can again conduct word study of word roots, affixes, synonyms/antonyms, and analogies. As was true of step one, this word study enables us to use basically the same materials (although modified) as for the regular classes. It is critical that ELLs have practice of this type for the end of the year standardized tests. Review and repetition at the beginning and end of a lesson gives them confidence that they can decipher an unfamiliar word they may encounter on a test.

Developing fluency. A number of strategies are used to develop fluency. Nation (2008) has demonstrated the exponential benefits of focusing time on fluency activities. He suggests including fluency activities as one-fourth of a program because it increases grammatical knowledge and skills. We will describe two different types of fluency activities, one for review of assignments and the other for review of concepts.

Fluency review takes place before the assignment is done and afterwards. Rather than handing out an assignment and perhaps giving a quick sentence or two of clarification, we take time to explicitly go over the directions, information, and analyze questions together. We essentially conduct a discussion about the purpose and expected outcome of the assignment. It provides students with directives for accomplishing it. Generally there a lot of questions from students at this point, and we do this exposition until the teacher is satisfied that the ELLs understand their task. When an assignment is due, for example, a fill in the blank, or a completion of sentence stems, teachers can quickly conduct an information gap check of their understanding. The grade given to students is simply for completion of the blanks or gaps in the assignment.

In step three, right after they have completed the assignment, we come back into a whole group and go over it together. We orally review and discuss the procedures and answers to the analysis questions. This is *not* simply the teacher telling them what the right answer is, but rather it is the teacher and the ELLs discussing the best ways to answer the questions and arriving at a consensus together and the reasons one answer is better than another answer, correcting misunderstandings, elaborating, explaining, etc. The ELLs check their own answers for accuracy and develop a deeper understanding of the content and skills required of them. Doing this is explicitly going over the concepts again. It is reiterative in that the teacher and the ELLs are going over the same content for a fourth or fifth time at this point, after having done so first in the whole-group examination of the assignment, then practicing in groups and answering the teacher's oral questions, answering the assignment's analysis questions in writing, and finally going over the concepts again when back together in a whole-group discussion of the assignment. It is appreciated, meaningful, timely feedback. This type of repetitive activity builds fluency and works for both teacher and student—the teacher does not need to spend hours grading assignments and making comments on all of them, and the students immediately correct their work while benefitting from the exposition.

Mnemonics. Building fluency requires familiarity with the subject. For ELLs, repetition is a key strategy and can be engaged in as an introduction, but most often once the language and content have been learned. An example of a fluency activity that we use to review concepts is to work together to create a mnemonic. The mnemonics can be to review content, or language as in the example that follows.

Ana teaches that the S in "SUMmary" stands for Some of the information, the A in "PARaphrase" stands for All the information, and the DEFine stands for Describe it. Afterwards, I have them take responsibility for the rest, such as the P and R in paraphrase could be, "Put into your own words" and "Reorganize." The U and M in summary could be, "Use key words and Main ideas." The E and F could be, "Examples, and Fits into a chapter." Once we have the mnemonics, we revisit the assignment. Does it require paraphrasing, summarizing, or defining? We might even call and respond, an oral strategy in which half the class says the word and the other half recites the mnemonic.

The result is ELLs understand the homework assignment, learn ubiquitous academic words, and a study technique as well (Fig. 9.2).

ELLs also need opportunities to improve their fluency with the information being learned. ELLs need repetition activities in which they develop accuracy and speed with the newly learned English. Some of the best materials for this are dramatic readings, poems, and songs. They are taught with such CLT strategies as recitation of jingles, poems, and choral reading. These are powerful tools for committing something to memory. They help them develop automaticity. In other words, the ELLs are able to process the English using the new vocabulary faster when they listen and read. In addition, by writing and saying the poems, jingles, and rhymes they improve expressive language skills.

There are several steps to conducting a fluency jingle in science. Here is one example Ana has used. First, Ana teaches them a simple and well-known song, such as Happy Birthday, or Skip to My Lou. Then, Ana employs a ubiquitous strategy for ensuring comprehension with ELLs, showing pictures and having them make personally relevant connections to the new vocabulary. The following shows ways one class studying clouds personalized the new words in making summaries and description of facts with few words (Fig. 9.3).

PARaphrase	SUMmarize	DEFine
Put it in your own words	Some information. Short	Describe it
All the information	Use key words	Examples
Reorganize	Main ideas	Fit it into a chapter topic

Fig. 9.2 Developing mnemonics for paraphrase, summary, and define

Cloud Facts

Latin roots: Alto - Latin altus for "high"

Cumulus - Latin word for "heap" or "pile"

Stratus - Latin word for "layer"

Facts:

Cumulus clouds are fluffy, puffy, and tall. In summer they slowly grow upward and look like cotton and cauliflower. They can make a thunderstorm. They block the sun.

Stratus clouds are grey and low. They cover the tops of tall towers and hills. They block the sun.

Fig. 9.3 Using word parts and simplified text

After they learn the song, and learn the meanings of the content, next, we *stop and study* in order to connect key words or parts of words to what they know, such as Cumulus clouds look like cotton. Then, when the word has a productive root or affix we connect that, such as cumulo- means "heap" or "pile" in Latin. They can look up words in the dictionary that also contain CUMU, like accumulation, cumulative, etc.

Then I model the rewriting of the lyrics of the song that we learned with one new science vocabulary word. For example, here are the lyrics to a song with the model of defining Cumulus clouds.

Lou Lou skip to my Lou,
Lou, Lou skip to my Lou,
Lou, Lou skip to my Lou.
Skip to my Lou my darling.
Cumulus, cumulus puffy pile,
Cumulus, cumulus fluffy heap,
Cumulus, cumulus cauliflower,
Cumulus looks like cotton.

Then they work in pairs to write subsequent lyrics about the same concept. For example

Cumulus, cumulus puffy white.
Cumulus, cumulus fluffy grey,
Cumulus, cumulus thunderstorm,
Cumulus floating cotton.

Then they work in pairs to write subsequent lyrics about the other new concept using the new vocabulary, Stratus.

Stratus, stratus low and grey.
Stratus, stratus block the sun
Stratus, stratus, low and grey
Stratus clouds mean no sun today.

Finally, the students place pictures of the clouds around the room. Each group then takes turns reciting their newly written jingle to the class. The class must point to the cloud they are describing.

In sum, we increase student participation through a three-step routine. First, the whole group is introduced to the language used with the science concepts. The focus of the instruction during this step is on key concepts. Teachers ignore or remove extraneous language and guide students to the key ideas and teach the fundamental skills by connecting oral language to written text. Second, the students are grouped for activities to practice and deeply process material cooperatively. By circulating among the groups the teacher interacts with each to personalize and individualize the language and science to the ELLs' needs. The third step involves the whole class again for reviewing, rehearsing, restating, and recasting. Fluency development for ELLs depends on creating opportunities to engage in repetitious activities that increase their processing speed and accuracy. The three steps to lesson implementation build successful participation, because they allow an increased use of instructional strategies based on ELLs' abilities, such as deliberate language and science instruction with simplification, explicit modeling, and connections between oral and written English.

Implications for Teacher Education

The difficulty of science for ELLs stems from more than just the fact that the science text is too difficult. ELLs are also experiencing an overload due to the fact that they are learning science while at the same time attempting to comprehend and produce oral and written language. In addition, ELLs are learning concepts which assume foundations they may not have.

In a regular class, the teacher can focus on science and may give three or four assignments associated with learning each concept. After a lecture, the students are largely expected to complete the assignments on their own, to ask for help if they need it, and to complete most of the work outside of class. The teacher also sets time aside from class time to grade their work. However, ELLs cannot be expected to work independently due to lack of science and language skill, reluctance to ask for help, difficulty understanding lectures, and difficulty comprehending the academic language of science texts.

In classes with ELLs, the teacher focuses on one concept and gives fewer assignments in order to allow for mastery of the concept and completion of the assignment in English. We want ELLs to be successful participants in science classes and rather than expect them to complete assignments on their own, we set aside time for reviewing and completing assignments together in class. Reviewing assignments creates instructional conversations with tellability (i.e., congruence in the language used by the teacher and the language understood by ELLs and that appears in the materials). The instructional conversations have synchronicities among the teacher, student, and contextual variables and in this way provide ELLs with the repetition and recursive language that they need.

We create instructional conversations with tellability by using CLT strategies and the concept of less is more. The CLT strategies are aimed at increasing their comprehension, expression, study, and fluency in academic oral and written English. The teacher models strategies for comprehending a text, for speaking and writing about science, for studying science concepts, and for developing speed and accuracy in using English.

The concept of less is more aims to increase success. The teacher carefully selects one assignment, simplifies the text, and models participation. It may take several days to complete one assignment in this way, but the ELLs accomplish more. The implementation of CLT strategies with less content allows for more learning. Less is more requires a different type of classroom routine than that used in regular classes.

We combine the teaching of science and language by using a three-step routine. Out of necessity, this three-step process of whole-group/small-group/whole-group instruction requires more time to implement as compared to the typical instructional routine of a mainstream class. Time is needed to use CLT strategies that require ELLs to spend time processing language, examining text, and engaging in rich interactions. Additional time is required of teachers at the front end because they are selecting the key concepts of the text, simplifying assignments, and identifying difficult aspects of texts.

However, one of the benefits of the three-step routine is that there is economy at the back end and in the use of patterned, manageable routines. There is less to grade when the teacher routinely asks the same essential questions throughout, formatively assesses progress, and reviews assignments with students in class. In fact, students are producing more work to the teacher's satisfaction.

If our goal is to engage ELLs successfully, then we must meet them where they are and not overload them with an overwhelming amount of science and language which caused frustration. We do this by using CLT strategies that minimally increase the teacher's workload at the front end. Teachers expend time at the front end to eliminate nonessential content, focus on main concepts and adding strategies for teaching language skills. Implementing CLT strategies reduces teacher work at the back end. There are fewer assignments and these are reviewed and corrected in class. At the same time, these strategies increase the participation of ELLs. CLT strategies are implemented in a three-step routine that accelerates their ability to learn English and science. Students experience success and gain confidence in their science skills, their ability to access academic texts, and their ability to express themselves in English.

References

August, D., & Gray, J. L., "Developing comprehension in English language learners: Research and promising practices," pp. 225–245 in K. Ganske & D. Fisher (Eds.), *Comprehension across the curriculum: Perspectives and practices K-12* (New York, NY: Guilford, 2010).

Burke, K., Hand, B., Poock, J., & Greenbowe, T. (2005). Using the science writing heuristic: Training chemistry teaching assistant. *Journal of College Science Teaching, 35*(1), 36–41.

Cloud, N., Genesee, F., & Hamayan, E. (2009). *Literacy instruction for English language learners: A teacher's guide to research-based practices*. Portsmouth, NH: Heinemann.

Cummins, J. (1981). The role of primary language development in promoting educational success for language minority students. In California State Department of Education Ed., *Schooling and language minority students: A theoretical framework* (pp. 3–49). Los Angeles: California State University, Evaluation, Dissemination and Assessment Center.

DaSilva Iddings, A., Risko, V., & Rampulla, M. (2009). When You Don't Speak Their Language: Guiding English-Language Learners through Conversations about Text. *The Reading Teacher, 63*(1), 52–61. Retrieved from http://www.jstor.org/stable/40347651.

Dreher, M., & Gray, J. (2009). Compare, contrast, comprehend: Using compare- contrast text structures with ELLs in K-3 classrooms. *The Reading Teacher, 63*(2), 132–141.

Echevarria, J., & Graves, A. (2003). *Sheltered content instruction: Teaching English-language learners with diverse abilities* (2nd ed.). Boston, MA: Allyn & Bacon.

Harris, T., & Hodges, R. (Eds.) (1995). *The literacy dictionary: The vocabulary of reading and writing.* Newark, DE: International Reading Association.

Herrell, A., & Jordan, M. (2012). *50 Strategies for teaching English language learners.* Boston, MA: Pearson.

Keys, C., Hand, B., Prain, V., & Collins, S. (1999). Using the Science Writing Heuristic as a tool for learning from laboratory investigations in secondary science. *Journal of Research in Science Teaching, 36*(10), 1065–1084.

Labov, W. (1972). *Sociolinguistic patterns.* University Park, PA: University of Pennsylvania Press.

Lee, O. (2005). Science education with English language learners: Synthesis and research agenda. *Review of Educational Research, 75*(4), 491–550.

Linik, J. (2004). Growing language through science. *Northwest Teacher, 5*(1), 6–9.

Mohr, K., & Mohr, E. (2007). Extending English-language learners' classroom interaction using the response protocol. *The Reading Teacher, 60*(5), 440–450.

Nation, P. (2008). *Teaching vocabulary: Strategies and techniques.* Boston, MA: Heinle ELT.

Nation, P., & Deweerdt, J. (2001). A defence of simplification. *Prospect, 16*(3), 55–65.

Nunley, K. (2004). *Layered curriculum: The practical guide for teachers with more than one student in their classroom* (2nd ed.). Amherst, NH: Brains.org.

Nutta, J., Bautista, N., & Butler, M. (2011). *Teaching science to English language learners.* New York, NY: Routledge.

Perez, B. (1996). Instructional conversations as opportunities for English language acquisition for culturally and linguistically diverse students. *Language Arts, 73*(3), 173–181.

Swain, M. (1995). Three functions of output in second language learning. In G. Cook & B. Seidlhofer (Eds.), *Principles and practice in the study of language.* Oxford, UK: Oxford University Press.

Ana Lado is a Professor of Education at Marymount University in Arlington, Virginia, USA, and directs the Professional Master's Degree in Education with Concentrations in TESOL, Special Education, Curriculum, and STE(A)M. Additionally, she trains teachers and develops materials for Lado International College in Washington, D.C., an intensive English school started by Dr. Robert Lado in 1972. She teaches English in thousands of non-elite schools in Pakistan via radio. This project won an international award for sustainable development and is supported by the U.S. State Department. She authored *Teaching Beginner ELLs with Picture Books: Tellability* (2012).

Adrienne Wright is a Teacher of Biology at Yorktown High School in Arlington, Virginia, USA, and teaches HILTEX Biology (High Intensity Language Training Exiting) (WIDA levels 3 and 4 mostly). She is a graduate student at Marymount University in Arlington, VA, USA, in the TESOL program.

10

Scaffolding Science Vocabulary for Middle School Newcomer ELLs

Yuliya Ardasheva

Language is critical to building knowledge in science as students are increasingly asked to read complex informational texts and to develop sophisticated means of oral and written expression, as expected by the Next Generation Science Standards [NGSS] (NGSS Lead States 2013). Developing literacy in content areas is particularly important for English language learners (ELLs), whose still-emergent academic English language skills often undermine their academic performance across subject areas (see discussion in Ardasheva et al. 2016).

Mastering the academic language necessarily involves the acquisition of "general academic and domain-specific" vocabulary, expected by the Common Core State Standards [CCSS] (National Governors Association Center for Best Practices and Council of Chief State School Officers 2010, p. 8). This is essential in academically demanding subject areas such as science where general academic words often take on

Y. Ardasheva (✉)
Department of Teaching and Learning, Washington State University Tri-Cities, 2710 Crimson Way, Richland, 99354 WA, United States
e-mail: yuliya.ardasheva@tricity.wsu.edu

© The Author(s) 2017

L.C. de Oliveira, K. Campbell Wilcox (eds.), *Teaching Science to English Language Learners*, DOI 10.1007/978-3-319-53594-4_10

science-specific meanings and where science-specific words are both essential for learning, serving to label new knowledge, and hard to grasp due to their abstract nature and low frequency outside of science readings (Cervetti et al. 2015). Acquiring such complex and wide-ranging vocabulary is particularly challenging for secondary school newcomer ELLs who have fewer years to master the language skills needed to fulfill academic requirements for graduation (Short and Boyson 2012).

Yet, few research-based resources on how to meet ELL vocabulary needs in upper grades science classrooms exist (August et al. 2016), thus undermining the ability of teacher preparation programs to adequately prepare educators to work with adolescent ELLs, particularly those ELLs who are new to the country. To address this gap, this chapter reports on a case study documenting effective vocabulary development practices of an experienced middle school science teacher collaborating on a larger science-literacy integration project.

Newcomer ELLs' Vocabulary Needs

Most "vulnerable" ELL population. Currently, approximately one in nine U.S. students is an ELL, with about one student in ten, in some states, having interrupted schooling (Advocates for Children 2010). The situation is even more complex for newcomer ELLs who are typically foreign-born and who—unlike native-born ELLs do not all start their English education in kindergarten, but sometimes in middle and high school years (Short and Boyson 2012). Not surprisingly, then, foreign-born ELLs are much more likely to have limited skills in English. In 2013, for example, over 80% of all people speaking English less than "very well" in the United States were foreign-born (Zong and Batalova 2015).

Newcomer programs, established specifically to address the many challenges faced by this growing student population, have the development of beginning English language and literacy skills as their primary instructional foci (Short and Boyson 2012). Yet, despite research suggesting that acquiring academic language proficiency may require four to ten years (Hakuta et al. 2000), newcomer program services are typically available to students for only one year. After one year, newcomers enter mainstream classrooms where most teachers are simply not prepared to teach initial literacy skills (Short and Boyson 2012), including vocabulary.

ELL vocabulary gap: Depth and breadth. Much research has documented a strong relationship between vocabulary knowledge and reading achievement across subject areas and among students of varied backgrounds (e.g., August and Shanahan 2006) and a substantial ELL gap in vocabulary (e.g., August et al. 2009; Proctor et al. 2005). Indeed, while a small proportion of new words may be tolerated by skilled readers (who can infer the meanings of new words from context; Carlo et al. 2004), high proportions of unknown words may be disruptive to comprehension. This makes ELL students—whose vocabulary gap is estimated at about two to three grade levels below grade norms (Proctor et al. 2005)—particularly vulnerable.

Further, the ELL word gap demonstrates itself not only in terms of *breadth*, the number of familiar words, but also in terms of *depth*, the extended word knowledge including literal meanings, morphological characteristics, and semantic associations (e.g., synonyms) (August et al. 2005). This is particularly problematic for two reasons. First, simply knowing a word's definition is not sufficient for effective use of that vocabulary in context (Nagy and Herman 1987). Second, both research and theory suggest that word knowledge and conceptual knowledge are co-developed, with some authors considering word knowledge to be part of the world knowledge beyond language skills (see discussion in Cervetti et al. 2015).

Dealing with such vocabulary gaps may be even harder for low-literacy newcomer ELLs. After all, even though the easiest way of accessing a word's meaning may be through translation (Nation 2001), this may not apply to domain-specific words such as "density" or "compression" as low-literacy ELLs may not have these words in their native languages (Miller 2009). This is especially true for ELLs with a history of interrupted schooling. Further, as noted earlier, newcomer ELLs have fewer years to master the language and academic skills needed to fulfill academic requirements for graduation (Short and Boyson 2012).

Science Vocabulary Demands

Types of science vocabulary. As reflected in current reform documents (CCSSI, 2010; NGSS Lead States 2013), many scholars (e.g., Quinn et al. 2013) identify two vocabulary types commonly encountered in science, namely, *domain-specific* and *general academic* words.

Domain-specific words—also known as technical vocabulary—refer to concept-loaded words such as "chromosome" (Harmon et al. 2005). Such technical words serve to "'name' our new knowledge" with greater precision and usefulness than common words, yet may be "harder to know and/or harder to learn" (Cervetti et al. 2015, p. 154). The latter may be attributed to a number of factors, including the highly abstract nature of these words, their low frequency outside of science readings, and their linguistic complexity (i. e., morphological complexity, Latin and Greek origins; Fazio and Gallagher 2014). These features set scientific vocabulary apart from everyday words. Yet, thorough understanding and retention of technical words is foundational for "the learning of subsequent concepts" (Harmon et al. 2005, p. 265) and for accessing dense, technical, and abstract readings in science (Fang 2008).

General academic vocabulary incorporates *procedural vocabulary*, words linking concept-loaded words such as "be the result of" and *nontechnical words*, not concept-loaded words such as "component" commonly encountered across academic disciplines (Harmon et al. 2005). According to Fazio and Gallagher (2014), the primary functions of academic words in science are to link technical scientific words and to "convey specific information about a science topic or investigation" (p. 1410). Indeed, it has been most recently argued that academic words may be only *supposedly* general, in reality taking on more restricted, discipline-based meanings when encountered in content areas such as science (Hyland and Tse 2007). Fisher and Blachowicz (2013) argued with an example: Even though the word "analyze" has shared basic meanings (to "separate a whole into its elements or component parts," p. 47), its application differs across disciplines. In language arts, students analyze ways in which authors develop themes using words to shape meanings and tone. In science, students analyze data and alternative explanations. This suggests that to support ELL science learning, teachers need to focus not only on science-specific vocabulary but also on teaching discipline-based meanings of general academic words, not assuming that "just because students may have encountered the word in language arts, this understanding will transfer" to science (Fisher and Blachowicz 2013, p. 47).

Further complicating the matter, the "supposedly 'small' words" (Fisher and Blachowicz 2013, p. 49)—beyond science-specific and academic vocabulary—may pose additional comprehension problems for ELLs. Such "small" words (also referred to as Tier I vocabulary; Beck et al. 2013) are the everyday words designating common concepts (e.g., "water," "rock"). Not knowing the meanings of such "small" words can impede science teaching and learning of low-literacy ELLs, as exemplified by a teacher quote from Miller et al. (2005), "[How] can you teach them ecology or evolution when they don't even know what is a male or a female, sperm or egg?" (p. 30). Yet, due to their high frequency in everyday conversations, these "small" words are not typically recommended for instruction (Beck et al. 2013; see Cervetti et al. 2015) or included in science vocabulary interventions for ELLs.

Science vocabulary interventions for ELLs. In ELL science-literacy interventions, vocabulary (science-specific and general academic) may be focused on either as a mediator of broader learning outcomes (as measured by student growth on science or reading tests) or as a valuable outcome in and of its own right (as measured by growth on vocabulary tests). While the first line of research has history in the emergent field of ELL science education (e.g., Lee et al. 2008; Shaw et al. 2014[1]), the second line of research is still very limited. August et al. (2016), in particular, both highlighted the above-mentioned gap—noting that among 11 ELL-focused experimental vocabulary interventions conducted between 1985 and 2013 only one (August et al. 2009) was

[1] Although vocabulary was assessed in this study, the intervention itself did not include an explicit vocabulary instruction component to qualify for the second line of research.

focused on science vocabulary—and contributed to filling the noted gap in the literature with a new study.

In their 2009 study of Grade 6 students (562 ELLs; 328 non-ELLs), August et al. examined the effectiveness of *Quality English and Science Teaching* (QuEST), a literacy-science integration intervention delivered by project science teachers. The QuEST's vocabulary instruction component (15 new words per week) included: the use of glossaries with visuals, simple definitions, and Spanish translations; word learning strategies (word parts, cognates) instruction; and the embedding of the target words in guided readings and discussions. After 9 weeks of the intervention, the vocabulary growth effect size favoring ELLs in the treatment condition was .26. Authors attributed this low-end impact to the low-level of program implementation by science teachers (at 2.2 points on a 4-point scale). This may be attributed to science teachers' often not seeing themselves as language educators relying instead on language education programs to teach students literacy skills and assuming that "students will transfer strategies learned in a general context to science learning" (Fazio and Gallagher 2014, p. 1410).

In a subsequent study, August et al. (2016) reported substantially larger vocabulary growth effects of .57 and 1.70 at the end of a shorter, 5-week summer program intervention. Notably, there was a high level of program implementation, which was delivered by language arts teachers (in alignment with what students were learning in their science classrooms). The study compared the effectiveness of *embedded* versus *extended* vocabulary instruction in a group of 509 Grade 3 and 4 ELLs. Embedded vocabulary instruction consisted of providing students with brief definitions of new words (these definitions appeared in project-designed texts, next to the targeted words). Extended vocabulary instruction included: (1) *direct instruction*, 16 words per week were pretaught during weekly discussions of science and language objectives using picture cards, sentence strips, and student glossary entries; (2) *multiple exposures to new words*, including in language arts (words were highlighted in project-developed texts and posted on word walls, students were instructed to listen for targeted words during interactive reading and assessed on their understandings at the end of the unit) as well as in science classroom; and (3) *reinforcement*, in groups, students played vocabulary games at the end of the week.[2] Notably, while the extended intervention effect size was almost three times larger than that of embedded instruction, both effects were statistically and practically meaningful.

None of the intervention studies in either line of research, however, explicitly focused on the needs of newcomer ELLs, and none of these studies focused on examining practices of experienced science teachers with extensive language support preparation. The latter, as suggested by the differences in fidelity of implementation in August et al.'s 2009 versus 2016 studies, may be an important factor in how science vocabulary instruction is carried out. Provided that the vocabulary needs of low-literacy, newcomer ELLs are more extreme than those of other student groups and that science literacy instruction is rarely provided by language arts teachers, learning about effective practices of experienced science educators with language support expertise in supporting newcomer ELLs needs is particularly important.

A Case Study of an Experienced Science Teacher's Practices

Case Study Background

This chapter reports on a case study of a middle school science teacher and his students participating in a larger science-literacy integration study, with Planetarium Visualizations (3-D data, videos, animations; delivered by a planetarium educator) providing science enhancement and the Science Vocabulary Support (SVS) program + reading days (described below; delivered by the science teacher) providing literacy enhancement. The science-literacy integration program was implemented in a newcomer school serving Grades 6–11 ELLs who are in their first year of U.S. school enrollment and who score below level 2

[2] August et al. (2016) did report, however, that the participating teachers dropped some of the vocabulary games due to insufficient time.

("beginning") on the district's English proficiency test. The program was delivered over a period of five months (some program activities were piloted prior to implementation) and, at the middle school level, was aligned with two of the school's science units (Earth Science and Astronomy). During the year of the program, the school served ELLs speaking over 25 primary languages. About 25% of the students had a history of interrupted formal schooling and about 75% were refugees.

The SVS program included introducing students to ten thematically related words—six new and four review words—each week. The words were explicitly taught on Monday (using a teacher PowerPoint including, per each slide, a new term, its definition, and visuals) and incorporated in daily 5–15 minute word study activities (e.g., picture match, spelling pyramid, cloze; for more details see Ardasheva and Tretter 2017). A Vocabulary Journal (sets of graphic organizers providing space for each new word, its definition, graphic representation, and notes; Marzano and Pickering 2005) was used throughout the program implementation. Reading days were delivered over a period of one unit (Earth Science) in 50-min sessions and involved students in paired-reading and summary writing (see Ardasheva et al. 2015a; SVS and reading days sample materials and classroom vignettes can be found at: http://louisville.edu/planetarium/research/implementation/literacy-support). Prior to program implementation, the teacher was introduced to relevant theory and directly experienced a 1-week worth of materials as a learner, but was not provided with any implementation scripts.

The participating teacher and his students. The teacher, a Caucasian male, has taught for 18 years (including three years in the newcomer school). The teacher was highly qualified in science and ELL education, holding an Early Adolescence Science Certification from National Board for Professional Teaching Standards and an English-as-a-Second-Language (ESL) endorsement—including coursework in ELL methods, linguistics, and ELL-focused practicums—from an accredited teacher preparation program. All ELLs in his three science classrooms (Grades 6, 7, and 8) participated in this study and experienced the same science curriculum. Because of the year-round enrollment, the number of participating students varied from 65 at the beginning of the study to 79 at the end of the study; student age ranged from 10 to 14 years old.

Tools for identifying effective vocabulary development practices. The ethnographic approach used in this case study was grounded in the authentic practice of the teacher, featuring ongoing and extensive researcher–teacher collaboration. This included curriculum planning, weekly observations, videotaping, post-instruction reflections, and ongoing informal conversations about the program, including email communications. Data were analyzed using inductive and recursive content analysis (Rossman and Rallis 2003) of field notes, interview transcripts, and online communication records. Similar observations occurring across different situations served as the basis for initial data categorization. Subsequently, these categories were subsumed under larger themes describing teacher practices and believes.

Vocabulary Development Practices of an Experienced Teacher

The interview and observation data analyses identified a wide range of vocabulary building strategies and two vocabulary instruction delivery approaches that the teacher used to support newcomer ELLs' vocabulary needs. The two vocabulary instruction delivery approaches were: (a) *incidental*,[3] a brief, "under-1-minute" application of often just a single, one-modality vocabulary development strategy and (b) *extended*, instruction delivered over extended periods of time using varied, multiple-learning-modality combinations of vocabulary development strategies. Most importantly, whereas the latter approach was applied to "glossing" both general academic vocabulary and "small," everyday words, the former approach was applied to teaching science-specific, technical words and phrases. As discussed below, balancing between these two approaches allowed the teacher maximize his impact on enhancing

[3] The term "incidental"—rather than the more often used term "embedded"—was chosen to better describe the teacher's practices. That is, although both terms suggest brevity in application, the term "embedded" refers to vocabulary instruction that consists of providing a brief verbal definition to a new word (August et al. 2016), not reflecting the wide-ranging repertoire of verbal and non-verbal strategies used by the teacher.

newcomer ELLs' vocabulary knowledge in terms of both breadth and depth, with minimum demands on instructional time.

Incidental vocabulary instruction. Describing his incidental vocabulary instruction techniques, the teacher elaborated:

> I don't teach many non-science words intentionally, [though] we do some smaller words, especially if there can be a kinesthetic way to demonstrate, or a tactile way to show. Of course, the more senses you can involve in learning, the better. [So, I also] like to scaffold vocabulary with images, simple definitions, and gestures.

Such "non-intentional," quickly-served strategies noted during observations are described below under two categories, namely, non-verbal and verbal means of word meaning representations (though, the distinction between the two was not always clear-cut). As illustrated in the examples below, the incidental application of strategies—requiring only a few seconds—allowed for targeting substantial numbers of new words within and across lessons, thus potentially contributing to developing newcomers' breadth of vocabulary knowledge.

Non-verbal means of word meaning representations. This category included gestures, enactment, and visuals. Over the course of the observations these strategies were ever-present and were used to either provide access to the word meanings directly or to "visualize" a verbal definition—simple or extended (e.g., through analogy)—as illustrated in the examples below.

During one lesson, the teacher illustrated the idea that rocks can have opposite qualities by, first, holding his hands opposite each other and then elaborating:

> "Hard" and "soft" . . . they are opposites. So, some rocks are hard [the teacher knocks on the hard surface of the desk]. Some rocks are soft [. . .], you can hold them in your hands and crush them [the teacher mimics a crushing motion with his palm].

To illustrate the meaning of the word "crack," that came up in the same reading about rock properties, the teacher picked up a pencil and broke it in half to make a cracking sound (which he also verbalized by saying "Crack!"). At a later point, the teacher had his students drag their fingertips over tabletops to illustrate (and make them say) "smooth" and then feel sandpaper to illustrate (and make them say) "rough." When commenting on the use of gestures and enactment, the teacher noted that these could help students build visual connections to the content and observed: "I use gestures, some created by me, some student-generated." The teacher's pulling a student seating in a chair to enact "pull" illustrates the former (when discussing gravity) and asking students to enact "statue" (when discussing uses of natural resources) illustrates the latter.

The observed use of visuals, in turn, included either simple, on-the-spot drawings by the teacher or the use of existing (online, books, videos) images and photos. When coming across the word "swirl," for example, the teacher drew a swirl on the white board, while saying, "That's a swirl," and reinforced the meaning yet with another gesture (in essence, drawing a "swirl" in the air). When introducing students to the term "sentence frame," the teacher quickly searched Google images to illustrate the meaning of the new word "frame," first locating a picture frame, then that of a car to provide a brief definition for the word (so, "frames serve to hold something") and to draw analogy ("a picture frame serves to hold a picture and the car frame serves to hold a car; similar, a sentence frame serves to hold a sentence together").

Verbal means of word meaning representations. This category included connecting to students' prior knowledge or experiences and word consciousness raising. Similar to practices reported in Zwiers (2007), the teacher's connecting new words to prior knowledge relied on using questioning, examples, analogies, and personifications. Examples of such strategies included: comparing land mass "pressure" during a lesson on sedimentary rock formation with body mass "pressure" that Jose (a pseudonym) would experience if all other students in the classroom sat on him; comparing the "breaking" of diamonds (one of the hardest materials on earth) with the "breaking" of teeth (one of the hardest human body parts); or asking students to make predictions about lunch using a crystal ball, when clarifying the meaning of "prediction."

Zwiers (2007) argued that personifications, defined as "stepping into the mindset of a person or object and then acting or talking appropriately," may be especially effective because not only do they make learning more relevant and fun, but they also allow "students to feel safe when answering, knowing that their such responses cannot be seen as wrong" (p. 106). This observation is exemplified in the classroom vignette below:

T: Rocks can even melt. OK, if you were a snowman, show me how would you looked like if you melted.
Ss: [Students show a movement of something slowly going to the ground while making a sound, "ooo-waaaahhh."]
T: You even have the sounds too—good! So, rocks do the same thing. But why would a snowman melt?
Ss: Because it's hot.
T: Because it's hot. So, why would rocks melt?
Ss: Because it's too hot?
T: Because it's too hot. Same thing.

The observed word consciousness raising strategies, in turn, primarily focused on word parts (as illustrater later on in the chapter). Other strategies in this category included the use of cognates and metacognitive conversations (e.g., word origins, semantic relationships, and word learning strategies). Examples of the latter set of strategies included: the teacher's eliciting the cognate for "marble," which "sounds very similar in Spanish" (students first supplied the word "bola," the teacher then continued on to pressure for "mármol"); pointing out that "rubi" comes from "red" and it looks red (pointing out at the corresponding picture in the book); or giving the following directions to students to set them up for an independent reading session: "You can use a dictionary or you can use the computers. You can look up the words now, or you can just write the new words on a piece of paper and look them up later. You can figure them out [from context], maybe." Modeling of word learning strategies (looking up definitions, think-alouds) and reminding or enabling the use of such strategies (having computer/dictionary stations set up) were ever-present in the classroom as were strategies focusing student attention on word parts, as illustrated below.

When coming across the word "predict," while discussing the lesson objectives for the day, the teacher quickly wrote the word on the board, underlined its parts, and wrote the meanings of the parts right under ("pre"/"before" and "dict"/"say") asking students to put the meanings of parts together to define "predict" ("before saying") and asking them to come up with some predictions of their own (as noted in an earlier "lunch" example). In another example, as the teacher talked about the need for ELL students to be "advocates" for themselves when they enroll in regular education schools, the teacher made a quick remark, "'Advocate' is the same word as 'lawyer' in other languages. 'Ad' means 'for' and 'voc' means 'to speak.'" The students' gradual internalization of word parts' meanings is illustrated in the following vignette, in which the class came across the word "connect" while discussing lesson objectives for the day:

T: Hmmmm "connect," "connect"... [the teacher pretends to be puzzled]?
Ss: "Put together."
T: Yes. Because?
Ss: [in a choir] "Co" means "together."

Overall, when asked why focusing on small words with newcomer ELLs, the teacher elaborated, "I wouldn't be able to break down [for instance] 'photosynthesis' into its components because [...] students might not know the words 'light' or even 'with' yet."

Extended vocabulary instruction. As noted earlier, three features were characteristic of the teacher's extended vocabulary instruction, namely: (a) focus on science-specific, technical vocabulary; (b) longer instructional time allocations; and (c) a simultaneous application of a number of multiple-modality (verbal, non-verbal) vocabulary development strategies affording newcomer ELLs with multiple ways of accessing and processing the meanings of new words. In addition, extended science vocabulary instruction incorporated "in-teaching" of enabling words (words that are not the current instructional targets, but belong to larger conceptual schemas and are essential for enabling the comprehension of the target science words). Two classroom vignettes below exemplify the teacher's extended vocabulary instruction practices. Although more time-demanding, these practices are more conductive to developing students' depth of disciplinary vocabulary knowledge and are warranted considering that teaching of the science vocabulary also involves teaching of the corresponding science concepts (Cervetti et al. 2015).

When introducing "dwarf star," the teacher first used the SVS pronunciation strategy called echo (students' repeating the term three times following teacher's model in a loud, then quieter, then very-quite voice). The teacher, then, led a discussion about the term as students made a new Vocabulary Journal entry (the term, its definition, and graphic representation):

1. T: There is an English word, "dwarf" [...]. Does anyone know what does that
2. mean?
3. Ss: Me. It means old people with white birds.
4. T: Not necessarily.
5. Ss: But did you see the Snow White? They all did!
6. T: Oh, oh! I see, *Snow White and the Seven Dwarfs*? Ok. Ok. What is the common
7. characteristic of the dwarfs besides being old?
8. Ss: They are short.
9. T: They are short. They are small. So, what do you think a dwarf star would be?
10. Ss: A small star.
11. T: A small star. [...] Here is a dwarf star, a very small red dwarf star, and over here
12. is the giant star. [While working with the PowerPoint slide, the teacher points to
13. each star on the graph as he describes them.] So [the teacher reads the definition
14. while underlying each word he reads], "any star of average or low luminosity, mass,

15. and size."
16. Ss: What is "luminosity"?
17. T: Let's see.... mmmm. Spanish speakers?
18. Ss: Luminoso?
19. T: Which means "bright."
20. T: So, whenever you see "lu" it will probably has something to do with light.

In this vignette, the teacher initiated the discussion by eliciting students' prior knowledge of the word. That is, he asked if anyone already knew the word "dwarf" (which they did, having seen *Snow White and the Seven Dwarfs*, even though their interpretation of the word was incorrect; line 3). This was followed by the teacher's clarifying the everyday meaning of the word and gradually building up for the introduction of the science-specific meaning of the term using questioning and analogy [lines 5–10]. The teacher then reinforced the association between the term and its referent using visual representation strategy (pointing at the dwarf star on the graph while saying the word) and provided verbal definition of the term while supporting students' making oral-written language connections (underlying each word of the definition as it is being read out loud; lines 12–15). While working through the definition, the teacher also made sure that students understood the word "luminosity" (enabling vocabulary) which the students marked as new. The teacher did this in three ways: by eliciting Spanish-speakers' cognate knowledge, by providing a simple English definition, and by pointing out a word part "lu" that may be associated with "light" in other words. The latter strategy both provided an additional clue to the word's meaning and engaged students' "practicing" their word consciousness.

Introducing students to the next target term ("giant star"), the teacher repeated some of the vocabulary development strategies noted earlier (i.e., questioning, eliciting prior knowledge, using analogy, connecting to visuals, focusing on enabling vocabulary) and supplemented those with additional strategies most appropriate for elaborating on the meaning of the new term. As illustrated in the classroom vignette below, these additional strategies (including when working with enabling words) were: the use of louder voice (represented with ALL capital letters in the transcript) to model correct usage and to mark important information about the concept (line 5); connecting to other (math) content (lines 1–6, 20–25); highlighting (lines 24–25); and reinforcing previous vocabulary (lines 1–6 ["mass"] and 25–28 ["luminosity"]). Notably, the earlier-described strategy of developing word consciousness included focusing on both word parts and synonymy (lines 6–7, 14–17). The gesturing strategy (lines 9–10), in turn, initially elicited misinterpretations of the intended meaning, but was subsequently clarified by the teacher (lines 14–16; 25).

1. T: And here, we see a blue-white supergiant, 150 solar masses [The teacher points at
2. the corresponding picture on the PowerPoint]. Which basically means . . . ? If
3. something is 150 solar masses, what would that mean?
4. Ss: It has 150 more mass than Sun.
5. T: That's right. It's 150 TIMES the mass of the Sun. So, our Sun has 1 solar mass;
6. this planet has 150 of those. "Super" . . . do you remember what "super" means? It
7. means the same as "hyper."
8. Ss: Superhero.
9. T: Superhero. It is not just a regular hero, it is a hero that is [gesturing to suggest
10. grandeur, pointing hands up].
11. Ss: Big, giant.
12. Ss: Can fly.
13. Ss: Strange.
14. T: So, "super" means above. So, you can have a regular hero [gesturing at the chest
15. level] and you can have a superhero [gesturing to show big biceps, then lifting hands

16. above the head], which is even better that a regular hero. So, going back to our
17. image, this star is not just a giant, it is a super-giant. It is bigger than a giant. [The
18. teacher uses the echo strategy and reads the definition in chunks] . . . "a
19. star with much larger radius" (chunk1). [Teacher reads underling each word.]
20. Do you know what "radius" is . . . from math?
21. Ss: No.
22. T: [The teacher silently draws a cycle, its center, and a radius on the board.]
23. Ss: Oh.
24. T: This is "radius" [highlights in red]. And this is a "diameter" [draws a diameter in
25. blue]. It's just bigger [gestures to show bigger]. . . . "and luminosity" (chunk 2).
26. [Teacher reads while underling each word he reads.] Luminosity means?
27. Ss: Light. Bright
28. T: Light. Bright. Good!

Overall, when asked why focusing on explicitly teaching science vocabulary to ELLs, the teacher noted, "Intentional teaching of vocabulary is important in science." Without such an intentional focus on vocabulary, "ELLs may just view science as play-time, if just given a tray of materials." The teacher further elaborated that with direct pre-teaching of science vocabulary there would be "safety and comfort in having structure and simplicity, especially at first."

Implications for Teacher Preparation

At the same time as educational reforms (e.g., CCSSI 2010; NGSS Lead States 2013) require that all students master the academic vocabulary of informational texts, teachers face the influx of ELL students with still-developing English literacy skills. Considering these challenges, examining instructional practices that may positively impact this important student population's language and literacy development in academically demanding subject areas such as science is essential for informing teacher preparation programs. The main implications of this study are the following.

First, this study's results suggest that science teachers may benefit from more extended language preparation. Unlike other studies reporting science teachers' resistance to integrating literacy strategies learned during typically short-duration in-service professional development (e.g., August et al. 2009; Lee et al. 2007), the present study examining practices of a science teacher with extended ELL preparation (i.e., ESL endorsement) documented the teacher's strong buy-in to providing students with literacy supports and the teacher's effective use of a wide range of vocabulary-building strategies. This finding is unique as the typical science teacher practices, particularly at the secondary level, rarely involve an explicit focus on teaching literacy skills (Fazio and Gallagher 2014; Miller 2009; Scott et al. 2003). Although the present case study involves only a single participant (and research in this area, overall, is still very limited), the present study's results are supported by preliminary findings from a large-scale project in California funded by National Science Foundation (Stoddart and Mosqueda 2015). That is, the results from this project suggest that extended ELL-focused science teacher preparation—including an explicit focus on teaching the language of science incorporated into science methods and practicum experiences—can significantly increase teacher knowledge and self-efficacy in providing literacy supports for ELLs. The project also found that the extended ELL-focused science teacher preparation generated significant gains in student science and literacy outcomes. These gains were substantially larger than those typically reported in short-duration in-service professional development intervention studies (Shaw et al. 2014).[4] Considering the growing need for all teachers, including science

[4] Although the reported lower-end effect sizes were comparable to in-service professional development interventions, the upper-end effects were substantially larger (.95 for science concepts and vocabulary in writing).

teachers, to be aware of and be able to instructionally address the language demands of their content areas to better support growing numbers of ELLs examining the effectiveness of different ELL-focused teacher preparation models is paramount. The results of this study suggest that requiring ESL certification of science teachers working with ELLs may be one viable option. The results of Shaw et al.'s (2014) study suggest that fusing literacy into science methods course-work may be another viable option. Another possibility, in need of additional research, is to incorporate an instructional focus on content area language demands (including those of science) into ESL teacher preparation programs. This may insure more productive content area-ESL specialist collaborations within the schools.

Second, the results of this study suggest that effective language preparation for teachers may benefit from raising teacher awareness of not only the discipline-specific vocabulary demands, but also of the specific vocabulary needs of ELLs, particularly of those ELLs who are new to the country, and of the effective ways of addressing these needs using differentiated instruction. Indeed, the teacher in the present study understood that—in addition to general academic words that take specific meanings in science and discipline-specific words that are new to most students (Quinn et al. 2013)—ELLs would benefit from an instructional focus on "small," everyday words playing an enabling role in the comprehension of the disciplinary vocabulary and concepts. Yet, such everyday words are rarely recommended for explicit instruction (see a discussion in Cervetti et al. 2015). Relatedly, and even more importantly, the teacher was able to effectively differentiate his instruction to support the varied ELL vocabulary needs with minimum demands on instructional time. As noted earlier, this was achieved by balancing between incidental and extended vocabulary instruction approaches. The results of this study suggest that when "glossing" large numbers of "smaller" (everyday and academic) words encountered in science, the incidental vocabulary instruction approach may be most appropriate, eventually contributing to ELLs' developing breadth of vocabulary knowledge. When teaching the much smaller numbers of "big" (science-specific) words, which also involves the teaching of the corresponding science concepts (Cervetti et al. 2015; see also

Ardasheva et al. 2015b), the extended vocabulary instruction approach may be most appropriate, eventually contributing to ELLs' developing depth of vocabulary knowledge. Importantly, this study provides teacher educators with a research-based argument that instructionally addressing extensive vocabulary needs of ELLs, particularly those of newcomers, is both necessary to engage the students and feasible, with only minimal demands on instructional time.

References

Advocates for Children of New York (2010). *Students with interrupted formal education: A challenge for the New York City public schools.* Retrieved from: http://www.advocatesforchildren.org/SIFE%20Paper%20final.pdf?pt=1

Ardasheva, Y., Bowden, J. O., Morrison, J. A., & Tretter, T. R. (2015a). Comic relief: Using comic and illustrated trade books to support science learning in first year English language learners. *Science Scope, 38*(6), 39–47.

Ardasheva, Y., Norton-Meier, L., & Hand, B. (2015b). Negotiation, embeddedness, and non-threatening learning environments as themes of science and language convergence for English language learners. *Studies in Science Education, 51*(2), 201–249.

Ardasheva, Y., Newcomer, S. N., Firestone, J. B., & Lamb, R. (2016). Mediation in the relationship among EL status, vocabulary, and science reading comprehension. *Journal of Educational Research.* Published online: http://www.tandfonline.com/doi/abs/10.1080/00220671.2016.1175407?journalCode=vjer20

Ardasheva, Y., & Tretter, T. R. (2017). Developing science-specific, technical vocabulary of high-school newcomer English learners. *International Journal of Bilingual Education and Bilingualism, 20*(3), 252–271.

August, D., Carlo, M., Dressler, C., & Snow, C. (2005). The critical role of vocabulary development for English language learners. *Learning Disabilities Research & Practice, 20*(1), 50–57.

August, D., Branum-Martin, L., Cardenas-Hagan, E., & Francis, D. J. (2009). The impact of an instructional intervention on the science and language learning of middle grade English language learners. *Journal of Research on Educational Effectiveness, 2*(4), 345–376.

August, D., Artzi, L., & Barr, C. (2016). Helping ELLs meet standards in English language arts and science: An intervention focused on academic vocabulary. *Reading & Writing Quarterly, 32*(4), 373–396.

August, D., & Shanahan, T. (Eds.) (2006). *Developing literacy in second-language learners: Report of the national literacy panel on language minority children and youth*. Mahwah, NJ: Erlbaum.

Beck, I., McKeown, M. G., & Kucan, L. (2013). *Bringing words to life: Robust vocabulary instruction* (2nd *ed.*). New York, NY: Guilford Press.

Carlo, M. S., August, D., McLaughlin, B., Snow, C. E., Dressler, C., Lippman, D. N., Lively, T. J., & White, C. E. (2004). Closing the gap: Addressing the vocabulary needs of English language learners in bilingual and mainstream classrooms. *Reading Research Quarterly, 39*(2), 188–215.

Cervetti, G. N., Hiebert, E. H., Pearson, P. D., & McClung, N. (2015). Factors that influence the difficulty of science words. *Journal of Literacy Research, 47*(2), 153–185.

Common Core State Standards Initiative [CCSSI]. (2010). Common core state standards for English language arts & literacy in history/social studies, science, and technical subjects. Retrieved from http://www.corestandards.org/

Fang, Z. (2008). Going beyond the fab five: Helping students cope with the unique linguistic challenges of expository reading in intermediate grades. *Journal of Adolescent & Adult Literacy, 51*(6), 476–487.

Fazio, X., & Gallagher, T. L. (2014). Morphological development levels of science content vocabulary: Implications for science-based texts in elementary classrooms. *International Journal of Science and Mathematics Education, 12*(6), 1407–1423.

Fisher, P. J., & Blachowicz, C. L. Z. (2013). A few words about math and science. *Educational Leadership, 71*(3), 46–51.

Hakuta, K., Butler, Y., & Witt, D. (2000). *How long does it take English learners to attain proficiency?* ((Policy Report 2000–1)). San Francisco, CA: University of California, Linguistic Minority Research Institute.

Harmon, J. M., Hedrick, W. B., & Wood, K. D. (2005). Research on vocabulary instruction in the content areas: Implications for struggling readers. *Reading & Writing Quarterly, 21*, 261–280.

Hyland, K., & Tse, P. (2007). Is there an "academic vocabulary"?. *TESOL Quarterly, 41*(2), 235–253.

Lee, O., Luykx, A., Buxton, C., & Shaver, A. (2007). The challenge of altering elementary school teachers' beliefs and practices regarding linguistic and cultural diversity in science instruction. *Journal of Research in Science Teaching, 44*(9), 1269–1291.

Lee, O., Deaktor, R., Enders, C., & Lambert, J. (2008). Impact of a multiyear professional development intervention on science achievement of culturally and linguistically diverse elementary students. *Journal of Research in Science Teaching, 45*(6), 726–747.

Marzano, R. J., & Pickering, D. J. (2005). *Building academic vocabulary: Teacher's manual.* Alexandria, VA: Association for Supervision and Curriculum Development.

Miller, J., Mitchell, J., & Brown, J. (2005). African refugees with interrupted schooling in the high school mainstream: Dilemmas for teachers and students. *Prospect Journal: An Australian Journal of TESOL, 20*(2), 19–33.

Miller, J. (2009). Teaching refugee learners with interrupted education in science: Vocabulary, literacy and pedagogy. *International Journal of Science Education, 31*(4), 571–592.

Nagy, W. E., & Herman, P. A. (1987). Breadth and depth of vocabulary knowledge: Implications for acquisition and instruction. In M. G. McKeown & M. E. Curtis (Eds.), *The nature of vocabulary acquisition* (pp. 19–35). Hillsdale, NJ: Erlbaum.

Nation, I. S. P. (2001). *Learning vocabulary in another language.* Cambridge UK: Cambridge University Press.

National Governors Association Center for Best Practices & Council of Chief State School Officers. (2010). *Common core state standards for English language arts & literacy in history/social studies, science, and technical subjects.* Retrieved from http://www.corestandards.org/.

NGSS Lead States. (2013). *Next Generation Science Standards.* Retrieved from http://www.nextgenscience.org/sites/ngss/files/NGSS%20DCI%20Combined%2011.6.13.pdf.

Proctor, C. P., Carlo, M., August, D., & Snow, C. (2005). Native Spanish-speaking children reading in English: Toward a model of comprehension. *Journal of Educational Psychology, 97,* 246–256.

Quinn, H., Lee, O., & Valdés, G. (2013). *Language demands and opportunities in relation to next generation science standards for English language learners: What teachers need to know.* Stanford, CA: Stanford University, Understanding Language Initiative.

Rossman, G., & Rallis, S. F. (2003). *Learning in the field: An introduction to qualitative research* (2nd ed.). Thousand Oaks, CA: Sage Publications.

Scott, J., Jamieson-Noel, D., & Asselin, M. (2003). Vocabulary instruction throughout the day in twenty three Canadian upper-elementary classrooms. *The Elementary School Journal, 103*(3), 269–286.
Shaw, J. M., Lyon, E. G., Stoddart, T., Mosqueda, E., & Menon, P. (2014). Improving science and literacy learning for English language learners: Evidence from a pre-service teacher preparation intervention. *Journal of Science Teacher Education, 25*(5), 621–643.
Short, D. J., & Boyson, B. A. (2012). *Helping newcomer students succeed in secondary schools and beyond.* Washington, DC: Center for Applied Linguistics.
Stoddart, T., & Mosqueda, E. (2015). Teaching science to English language learners: A study of preservice teacher preparation. *Teacher Education & Practice, 28*(2/3), 269–286.
Zong, J., & Batalova, J. (2015). *The limited English proficient population in the United States.* Washington, DC: Migration Policy Institute. Retrieved from http://www.migrationpolicy.org/article/limited-english-proficient-population-united-states.
Zwiers, J. (2007). Teacher practices and perspectives for developing academic language. *International Journal of Applied Linguistics, 17*(1), 93–116.

Yuliya Ardasheva is Assistant Professor in ESL/Bilingual Education at Washington State University (WSU) Tri-Cities. Her research focuses on the interplay between second language and academic development and the contributions of individual differences to second language and academic development. She has led three grant-funded projects focusing on enhancing teacher practices in integrating language, literacy, and content instruction for ELLs. Dr. Ardasheva's work with diverse learners has been recently recognized with the WSU Martin Luther King Jr. Distinguished Service Award. She has published her work in numerous venues, including in *TESOL Quarterly, Language Learning,* and *Studies in Science Education.*

Index

© The Author(s) 2017

L.C. de Oliveira, K. Campbell Wilcox (eds.), *Teaching Science to English Language Learners*, DOI 10.1007/978-3-319-53594-4

C

D

E

CPSIA information can be obtained
at www.ICGtesting.com
Printed in the USA
BVOW06*0825131017
497591BV00004B/14/P